Wildlife Conservation and Management

Wildlife Conservation and Management

Edited by:

Ashok Kumar Bishoyi

Delve PUBLISHING

www.delvepublishing.com

Wildlife Conservation and Management

Ashok Kumar Bishoyi

Delve Publishing
224 Shoreacres Road
Burlington, ON L7L 2H2
Canada
www.delvepublishing.com
Email: orders@arclereducation.com

ISBN: 978-1-77407-798-6 (Hardcover)

ABOUT THE EDITOR

Dr. Ashok Kumar Bishoyi, is presently working as Assistant Professor (Microbiology) at Marwadi University, Rajkot, Gujarat, India. Prior to academic carrier, he has worked as scientist B at Gujarat Biotechnology Research Centre, department of Science and Technology, Govt. of Gujarat. During his scientific carrier, he has executed different research projects related to transcriptomics, genomics, plant tissue culture, animal tissue culture, virology, molecular marker development, etc. He has also successfully completed different external/ internal funded research projects. He graduated from Charotar University of Science and Technology, with doctorate in Biotechnology in the year 2017 and master graduate in Biotechnology from Berhampur university, Odisha. He has published number of research articles in different reputed peer reviewed international and national journals for his research credit. He has also reviewed number of research articles/books for different reputed international scientific journals/ publishers.

TABLE OF CONTENTS

GLOSSARY OF TERMS

A

Aesthetically – In a way that gives pleasure through beauty.

Anthropologists – An expert in or student of anthropology.

B

Bobwhite – A New World quail with mottled reddish-brown plumage, and typically a pale throat and eye stripe.

C

Chlamydial Infection – Is a sexually transmitted infection that may not present any noticeable symptoms.

Cryophilic Bacteria – Or cryophiles are extremophilic organisms that are capable of growth and reproduction in low temperatures, ranging from −20°C to +10°C.

D

Demographic – A particular sector of population.

Deoxyribonucleic acid – It is a long molecule that comprise a unique genetic code.

Deterioration – The process of becoming progressively worse.

Disseminated – Spread (something, especially information) widely.

Distemper viruses – Is a viral disease that affects a wide variety of animal families, including domestic and wild species of dogs, coyotes, foxes, pandas, wolves, ferrets, skunks, raccoons, and large cats, as well as pinnipeds, some primates, and a variety of other species.

Diversity – A range of different or variety of things.

E

Enforcement – The act of compelling observance of or compliance with a law, rule, or obligation.

G

Galvanized – Shock or excite (someone) into taking action.

H

Herds – A large group of animals, especially hoofed mammals, that live together or are kept together as livestock.

Hibernating – (of an animal or plant) spend the winter in a dormant state.

Hindrances – a thing that provides resistance, delay, or obstruction to something or someone.

I

Immunosuppressive – Is a reduction of the activation or efficacy of the immune system.

Inbreeding – Production of offspring from breeding or mating of individuals or organisms that are genetically related.

Intercept – Obstruct (someone or something) so as to prevent them from continuing to a destination.

Interconnected –Having all constituent parts linked or connected.

Invasive – Tending to spread very quickly and undesirably or harmfully.

M

Meagre – (of something provided or available) lacking in quantity or quality.

Microsatellites – It is a tract of repetitive DNA in which DNA motifs are repeated. Usually 5 to 50 times.

Mitochondria – Is a semi-autonomous double-membrane-bound organelle found in most eukaryotic organisms.

Monitoring – Observing the quality or progress of anything over a certain period of time.

Morphology – Study of particular form, shape, and structure.

Mortality – The state of being subject to death.

Moth – An insect with two pairs of broad wings covered in microscopic scales, typically drably colored and held flat when at rest. Moths are chiefly nocturnal, and lack the clubbed antennae of butterflies.

Muskrats – A large semi-aquatic North American rodent with a musky smell, valued for its fur.

N

Natality – The ratio of the number of births to the size of the population; birth rate.

O

Outbreak – a sudden occurrence of something unwelcome, such as war or disease.

P

Population dynamics – A branch of life sciences which studies the age and size composition of specific populations as dynamical systems.

Predation – The preying of one animal on others

Predators – An animal that naturally preys on others.

Proximity – Nearness in space, time, or relationship.

Pyramids – A monumental structure with a square or triangular base and sloping sides that meet in a point at the top, especially one built of stone as a royal tomb in ancient Egypt.

Q

Quail – a small short-tailed Old-World game bird resembling a tiny partridge, typically having brown camouflaged plumage.

R

Reproduction – A biological process by which new organisms are produced that are known as offspring. It is a principal feature of all known life.

S

Sediments – Matter that settles to the bottom of a liquid; dregs

Snags – An unexpected or hidden obstacle or drawback.

Starvation – Suffering or death caused by lack of food.

Stockpile – A large accumulated stock of goods or materials, especially one held in reserve for use at a time of shortage or another emergency.

Subsistence – The action or fact of maintaining or supporting oneself, especially at a minimal level.

T

Thylacine – Now extinct, is one of the largest known carnivorous marsupials, evolving about 4 million years ago.

Translocation – In general, it means a change in location. When a part of chromosome is transferred to another chromosome.

Transmission – The action or process of transmitting something or the state of being transmitted.

Triggering – (of an event, circumstance, etc.) causing a particular action, process, or situation to happen.

V

Veterinary – Relating to the diseases, injuries, and treatment of farm and domestic animals.

LIST OF FIGURES

LIST OF ABBREVIATIONS

AEWA	African-Eurasian migratory water birds agreement
ATP	Adenosine Triphosphate
BSCP	Bangladesh spoon-billed sandpiper conservation project
CBC	Community-Based Conservation
CBD	Convention Of Biological Diversity
CBFM	Community-Based Forest Management
CBNRM	Community-Based Natural Resource Management
CITES	Convention on International Trade in Endangered Species
CT	Conservation International
CTVT	Canine Transmissible Venereal Tumor
DFTD	Devil Facial Tumor Disease
DNA	Deoxyribonucleic Acid
EDGE	Evolutionarily Distinct And Globally Endangered
EU	European Unions
FAOSTAT	Food and Agriculture Organization Corporate Statistical Database
FMD	Foot And Mouth Disease
FRA	Forest Resources Assessment
GPS	Global Positioning System
ICDPs	Integrated Conservation and Development Projects
ICRW	International Convention for the Regulation of Whaling
IUCN	International Union for Conservation of Nature
JFM	Joint Forest Management
KoRV	Kola Retrovirus
LPI	Living Planet Index
MANOVA	Multivariate Analysis Of Variance
NAPAP	National Acid Precipitation Assessment Program
NGOs	Non-Governmental Organizations
NGS	Non-Invasive Genetic Sampling

PAs	Protected Areas
PCA	Principal Components Analysis
RNA	Ribonucleic Acid
SASUSG	Southern Africa Sustainable Use Specialist Group
SBR	Similipal Biosphere Reserve
SCB	Society for Conservation Biology
SNPs	Single Nucleotide Polymorphisms
TANAPA	Tanzania National Parks Authority
TEDs	Turtle Excluder Devices
UNEP	United Nations Environment Programme
USFWS	United States Fish and Wildlife
VCGs	Village Conservation Groups
WMPs	Wildlife Management Practices
WWF	World Wildlife Fund
ZSL	Zoological Society of London

PREFACE

Wildlife is an important part of the ecosystem we live in. It completes the various cycles of the ecosystem and contributes towards maintaining a healthy balance in the ecosystem. Wildlife comprises of various animals that cannot be or are not domesticated.

In recent times, however, in spite of the great importance that they hold, the wildlife has had a major hit. It has been found that many species have turned extinct or are on the verge of getting extinct. This is a huge loss of biodiversity and can prove detrimental for the survival of other dependent species as well.

For this reason, wildlife conservation has been an area of deep focus for the conservationists and the authorities and the stakeholders have been putting all their efforts to make sure that the management of wildlife is done in the most efficient ways.

This book brings the focus of the readers to the conservation and management of wildlife, elaborating about the various measures that have can help in the conservation practices and the actions that have been taken to ensure an increased wildlife diversity.

The book starts off by introducing the wildlife ecology to the readers. It informs them about the various kinds of biomes and gives them knowledge on the nutrition required by the wildlife, their shelter needs and the adaptation they have for the changing climate.

The second chapter brings the readers to the subject of population fluctuations in wildlife. It explains to them the life tables and the rate at which the population grows. It also informs about the basic components that play a role in the population change. The third chapter narrates the factors that are responsible for the declining wildlife populations and relates them to the recent cases that have been found to exist.

The fourth chapter focuses on the assessment of the population of the wildlife and the evaluation of the methods that must be used to conserve the depleting wildlife. There have been various approaches that have been taken to handle wildlife conservation effectively. The fifth chapter brings to light, these conservatory approaches that are taken from the theoretical and legal point of view.

The sixth chapter takes the readers through the practical aspects of the management of wildlife. The seventh chapter explains the readers the impact of the various kinds of practices that aim for wildlife conservation. These impacts have been observed on a wide range of aspects.

The eighth chapter talks about the future aspects of wildlife conservation and focuses primarily on the integration of the different kinds of conservation practices that may help in getting the wildlife scenario back on track.

This book is mainly to educate the readers about the worrying and worsening scenario of wildlife in our ecosystem and to inform them about the measures that are being taken up or must be taken up to ensure that the conservation processes are well aligned with the objectives and that the management of these practices is carried out in the most appropriate ways.

1

Wildlife Ecology: An Introduction

CONTENTS

Wildlife plays an important role in maintaining balance the nature. It provides stability to different environmental processes. Over the years the humans have modified the global landscape to accommodate the growing population and to meet their ever-growing needs for food and economic development, this has led to various activities like urbanization, extraction of resources, energy production that have drastic changes in our ecosystem.

As a result, humans have changed the environment in which wildlife thrived, like their food (both in quantity and quality). This had implications on animal health, reproduction, and survival. Moreover, the climatic changes brought about by shrinking of forests have led to disturbing the balance of wildlife at a location in various ways.

Wildlife Ecology is a field of study that dwells upon conservation and management of wildlife on ecological principles for maintaining biodiversity. In this chapter, various aspects of wildlife ecology and the researches being made for the conservation of various species of wildlife.

1.1. INTRODUCTION

1. **Wildlife** broadly refers to all wild plants, wild animals and other living organisms that live and grow independently in their natural habitat, without being introduced by humans. Wildlife is found in all ecosystems like deserts, forests, plains, grasslands, and even most developed urban areas. But distinct forms of wildlife exist in different ecosystems.

Figure 1.1. Wildlife forms a fundamental part of our ecology.

Source: Image by Pixabay.

Term wildlife in generally speaking terms refers to animals untouched by human factors, although it is a matter of fact, as argued by experts, that wildlife existing around the world is affected by human activities.

Historically, humans have tried to separate wildlife from civilization in moral, social, and legal sense. There is a mention of certain animals being revered or declared sacred in certain religions. In recent times, concern for the natural environment has led the scientists and experts to protest against the exploitation of wildlife and ecosystems for the benefits of humans only. Also, this awareness has resulted in the formulation of a number of laws for the conservation of natural environment as well as wildlife.

2. **Ecology** is the study of how living organisms including humans; interact with each other and with their physical environment. It seeks to understand the vital connections between plants and animals and the world around them.
3. **Wildlife Ecology** is the study of wildlife management that seeks to manage wildlife populations for the benefit of humans. Although, people hunt animals for food and fur, and also enjoy viewing wildlife, often the conflict situations arise because wild animals kill livestock, damage crops and cause collision of vehicles.

Wildlife ecology has progressively become more quantitative, even then it still relies on techniques, with emphasis on statistical methods rather than on ecological principles and broader concerns of preservation of biodiversity. As wildlife ecology was slow to embrace the modern ecological theory, a new subject namely Conservation Biology emerged in the 1980s. Since then, however, wildlife ecology has converged as a sub-discipline of conservation biology that largely focuses on the applied ecology and management of wild populations of birds and animals.

1.2. ECOSYSTEMS AND BIOMES

1.2.1. Ecosystems

An Ecosystem is a geographic area, where plants, animals, and other living organisms, as well as weather and landscape, work together to form a bubble of life. Ecosystems comprise of biotic or living parts and a biotic or non-living parts. Biotic parts or factors are plants, animals, and other living organisms and a biotic parts or factors can be summarized as climate, minerals, humidity, sunshine, and type of soil, etc.

The living and physical components of an ecosystem are therefore linked through nutrient cycle and energy flows. Each and every factor of an ecosystem are inter-dependent on each other. For example, change in temperature of an ecosystem often affect what types of plants will grow there and whether herbivorous animals, which are dependent on plants for their food and shelter, will adapt to such changes or move on to another ecosystem or perish.

Ecosystems can be very large or very small. The entire earth's surface is a series of ecosystems. Ecosystems are often connected to larger biomes.

1.2.2. Biomes

A biome is a very large geographical area, comprising of community of plants, animals, and other organisms that have common characteristics for the environment in which it exists. Biomes can be found over a range of continents and are distinct biological communities, in response to shared physical climate.

A biome is not an ecosystem, although it can look like a massive ecosystem. One may find many ecosystems within one biome. There are mainly two types of biomes, that are land based and water based. Land based biomes are known as Terrestrial biomes and water-based biomes are called Aquatic biomes.

Further, there are various types of Terrestrial biomes and Aquatic biomes, which are explained hereafter.

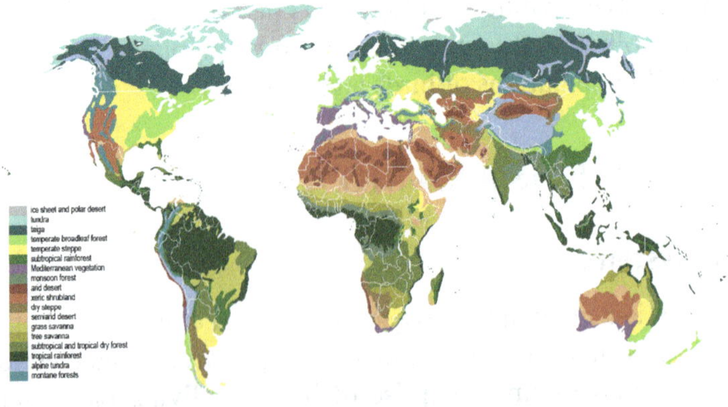

Figure 1.2. Different kinds of biomes on earth.

Source: Image by Wikipedia.

1.3. TERRESTRIAL BIOMES

1.3.1. Desert Biome

Deserts make up 20% of the land cover on earth. Desert biomes receive less than 12 inches of rainfall annually or no rainfall at all. Desert biomes are of four kinds that is, Hot, and Dry deserts, semi-arid deserts, coastal deserts and cold deserts.

Desert biomes experience very high temperatures due to less vegetative cover, less cloud cover and land's exposure to the sun. Plants that grow and survive are adapted to low rainfall and are short shrubs and cacti, that have the ability to conserve water. Examples of plants that grow in desert biome are yuccas and cacti.

Animals in desert biome tend to burrow or stay in a hideaway till sunset to avoid the heat. Some of the desert species of animals are reptiles, small carnivorous birds and insects that are adapted to survive with very little water.

Some examples of desert biomes are Sahara Desert of North Africa and Mojave Desert of California, USA.

1.3.2. Chaparral Biome

Chaparral biomes are characterized by shrub land and few trees. It has mild wet winters and hot dry summers. Chaparral biomes receive 15–39 inches of rainfall annually, mainly in winters. Dry summers are characterized by wildfires. Chaparral biomes are found in southern California and Baja California Peninsula, Mexico.

Chaparral biome is often confused with desert biome, as they share many similarities. Both biomes have hot and dry climate, but Chaparral receives more rainfall than desert biome.

Figure 1.3. Chaparral biome.

Source: Wikimedia Commons.

1.3.3. Tundra Biome

Tundra biome is the coldest terrestrial biome with the least biodiversity capacity. It is characterized by freezing temperatures and receives very little rainfall and it covers one fifth of the earth's surface. Tundra biome is located in high Arctic region around the North Pole and has a growing season of 60 days only.

Plants in arctic tundra are short and grow closer. Examples of plants found are mostly shrubs, lichens, mosses, and liverworts. Animal life in Tundra biome comprises of lemmings, caribou, migratory birds, fish etc.

1.3.4. Taiga Biome

Taiga biome extends south from Arctic Circle up to Temperate forests. Taigas are referred to as boreal forests or snow forests. Alaska, Canada, Scandinavia, and Siberia have Taiga biome. Taiga has long dry winters, cool, and wet summers and have a growing season of around 130 days. Annual precipitation in Taiga biome ranges from 16–40 inches, mainly in form of snow.

The plant life in Taiga biome comprises of coniferous trees and low plants. Animal life in Taiga consists of bears, lynx, moose, deer, hares, and woodpeckers etc.

Figure 1.4. Taiga biome.

Source: Image by Flickr.

1.3.5. Grassland Biome

Grassland biome mainly comprises of grass, as the name suggests and a few sparsely distributed trees. The two main types of grassland biomes are ***Savanna Grasslands*** and ***Temperate Grasslands.***

One major *savanna* is located in Africa and covers almost one-third of the African continent. Others are located in parts of India, Australia, and South America. Savannas receive concentrated rainfall for several months followed by spells of drought. The soil layer in savannas is thin and does not have the capacity to retain water. Animal life in savannas comprises of large mammals like lions, hyenas, giraffes, and buffaloes etc.

Temperate Grasslands are found in South Africa, Argentina, and some plains of central North America. These include steppes, veldts, and prairies. Moderate rainfall, rich soils, hot summers and cold winters are characteristics of this biome. Few trees grow near the rivers and the animal life includes deer, gazelles, birds, insects, and large predators like wolves and lions.

1.3.6. Rainforest Biome

Forests account for 30% of the total land cover on earth. They are of incredible value to life and are source of many raw materials on which humans depend. The rainforest biomes are of two types, that is, ***Tropical Rainforest Biome*** and ***Temperate Rainforest Biome.***

1. ***Tropical Rainforests Biome*** comprises of world's greatest biodiversity. They are located near the equator, experience warm temperatures throughout the year, have extremely high humidity and get more than 200 cm of rainfall annually.

Plants that grow in tropical rainforests are usually huge trees with buttress roots, lots of large green leaves that provide dense cover to the land. Epiphytic plants grow on trees and other vegetation. Amazon rainforest is an excellent example of a tropical rainforest biome.

2. ***Temperate Rainforest Biomes*** are found on higher altitudes, have cooler temperatures but significant precipitation. Plant life in temperate rainforests comprises of ferns, evergreen trees and mosses. The Olympic National Park of Washington State has temperate rainforests.

1.3.7. Temperate Deciduous Forest Biome

Temperate deciduous forest biome is ever changing. It has four distinct seasons i.e., spring, summer, autumn, and winter. Temperate deciduous forests get between 30 to 60 inches of precipitation all-round the year. The varied temperatures yield a diverse biome.

The plant life consists of deciduous broadleaf trees, evergreens amongst others. This biome has many animal species like deer, rabbits, birds, and amphibians amongst many.

1.3.8. Alpine Biome

This biome exists at high altitudes. This is a mountainous biome and usually at these levels' trees do not grow. Alpine biome has around 180 days of growing season. In alpine biome, the plant life contains a number of shrubs, grasses, and heaths. Mammals such as goat, sheep, elks amongst others exist in this biome. The high Sierra Nevada mountain range in California is an example of alpine biome.

Figure 1.5. Alpine biome.

Source: Image by Wikipedia.

1.4. AQUATIC BIOMES

Aquatic biomes pertain to water bodies. Mainly, aquatic biomes are of two types; Freshwater Biomes and Marine Biomes. Together, the Aquatic biomes make up around 73% of the earth's surface.

1.4.1. Freshwater Biomes

Freshwater biomes are water bodies in which salt concentrations are very low. Lakes, rivers, streams, ponds, and wetlands are freshwater biomes.

Thermal mixing takes place in lakes and ponds. *Rivers and streams* either move towards oceans and lakes or in some cases, lakes are the source of rivers and streams. Freshwater biomes are home to algae, fish, waterfowl, crustaceans, and many micro-organisms. The best example of a freshwater biome is Amazon river in South America.

Shallow water bodies like swamps, marshes, and bogs can be categorized as *Wetland Biomes*. In wetlands, water flow is slow and steady and these are hosts for many types of plant and animal life. Bolsa Chica Ecological Reserve in Orange County of California, USA is a typical example of a wetland biome.

1.4.2. Marine Biomes

Marine biomes are water bodies characterized by saltwater. Oceans, coral reefs and estuaries are examples of marine biomes. Abiotic factors like sunlight penetration in various layers and zones of the ocean lead to more biodiversity.

The zone near the shoreline is affected by tides and waves and known as ***intertidal zone***. The zone that extends to the continental shelf is called ***neritic zone***. Enough sunlight penetration for photosynthesis is there and this zone is characterized by presence of seaweeds. The ***oceanic or pelagic zone*** than extends farther, is home to large fish and mammals.

The deep zone beyond continental shelf, known as ***benthic zone*** is characterized bypresence with sponges and multiple varieties of sea fish. The deepest zone of the ocean that is ***abyssal zone*** is characterized by high pressure, absence of sunlight and cold temperatures.

1. **Coral Reef Biome** is found in shallow parts of the ocean in tropical zone. Coral reefs are made up of calcified remains of coral animals and are home to many underwater species. An appropriate example of coral reef biome is the Great Barrier Reef of Australia.
2. **Estuary Biomes** are found where rivers or streams enter the ocean. Crustaceans, waterfowls, and halophytic plants (plants tolerating salinity changes) are found in estuary biomes.

Figure 1.6. Estuary biome.

Source: Image by Wikipedia.

1.5. STUDY OF WILDLIFE ECOLOGY

As mentioned earlier, wildlife ecology is a sub-branch of conservation biology, which is focused largely on applied ecology and management of the population of birds and animals. Studies in wildlife ecology also involve the species that are hunted, species that are of importance to humans from either economic or aesthetic point of view including the rare or endangered species.

Though the thought for the development of modern environmental ethics and wildlife preservation had found mention in the works of experts earlier, but since inception, this field had become more and more quantitative. But modern wildlife ecology not only applies the ecological theory to management and conservation of animals, it also includes the theoretical studies on wildlife.

The focus in the field of wildlife ecology is based mainly on the population of a species rather than the individual, although population response can be termed as a sum of individual responses. As a consequence, ***population ecology*** has become the core of the conservation and management of wildlife.

Quantitative techniques are more often used in wildlife studies and with this, the significance of scale in understanding and analyzing ecological phenomena is becoming more and more evident. Substantial developments have taken place in this field like (a) assessment of population density or population size through surveys, (b) assessment of population parameters from capture-recapture methods and (c) estimation of occupancy and modeling.

In model building for analyzing population curves, survival, and reproduction, the information-theoretic approach has become more prominent. Population viability analysis and extinction analysis have become critical elements for studying wildlife populations, especially in case of rare or endangered species.

Apart from above, translocation of wildlife for re-establishing locally eradicated population of a species has become a standard technique of wildlife management, although the concerns are raised over establishing non-native species in a location. The role of genetics has assumed increasing importance in wildlife conservation, including investigating genetic structures amongst the population of a particular species.

The role of meta-populations and source-sink dynamics in conservation of wildlife, and in quantitative methods to deal with conservation issues has evoked the interest of ecologists.

Meta-population is a group of populations that are separated by space, but consist of the same species. These spatially separated populations interact as individual members move from one population to another.

Source-Sink Dynamics is a theoretical model used by ecologists to describe and determine how variation in habitat quality may affect the population growth or decline of organisms.

Density dependence in population ecology pertains to the effect of the present and/or past population on the per capita population growth rate. Density dependent processes take place when population growth rates are regulated by population density.

It is important to understand, whether a particular species exhibits a strong density dependence with increasing population size is a pre-requirement for the application of an appropriate population model or management strategy.

Nonetheless, large mammals typically display density dependence and certain aspects of their ecology or life characteristics can be correctly interpreted without going into the nature of density dependent processes.

However, in certain species, density dependent processes affect some of their aspects like size of their populations and variations therein, relations amongst the abundance of food and quality, physical condition and reproduction, interactions of population size with severe weather, patterns of migration, predator-prey dynamics and characteristics of meta-populations. Such outcomes are significant for understanding the allowable growth of populations.

1.5.1. Software Based Studies of Wildlife Ecology

Advances in use of global positioning system (GPS) and analysis of data obtained from radio-tracking of animals have a key role to play in wildlife studies. A number of public domain software like HRT and BLOSSOM offer techniques for studies of wildlife ecology.

Calculation of home-range size is mostly done based on kernel models. Kernel models of home range make an assumption of independence of home-range locations. Techniques for assessing autocorrelation were offered by Swihart, Slade, and Solow, which if left unadjusted, result in underestimation of home range size.

Analysis of patterns of animal movements has become an area of intensive research. BLOSSOM software has also been used to assess spatial distributions of predators and prey.

FRAGSTATS software includes a number of landscape metrics and helps in the study of landscape ecology by furthering the understanding of the relationship between wildlife and environments they inhabit.

For survival analysis Kaplan-Meier procedure is used commonly. Program MARK, apart from accommodating staggered entry of animals, also offers survival estimation under various types of data that include band recovery, nest survival and known fates.

For estimation of time of death, Failure rate models with parametric techniques are used, which allow examination of interactions and covariates, provided assumptions for parametric techniques are met. Also, for the determination of rates for cause specific mortality in wildlife populations, proportional hazards models (Programs MICROMORT or STATA) can also be used.

Further, for assessment of timings of life-history events, survival analyses can be used along with other new available methods. For the study of habitat selection and use by wildlife, data obtained through radio telemetry can be utilized.

Modeling approaches have nowadays taken place of simplistic analytical techniques of habitat data, as these are more appropriate to estimate the influence of both categorical and continuous variables on space use patterns by animals.

The sexes of many wildlife species, segregate spatially for most of the period annually. Multivariate analysis of variance (MANOVA) and principal components analysis (PCA) are used for the measurement of sexual segregation. MANOVA and PCA allow examination of variables at the same time. The use of multivariate procedures like PCA has become very common in studies of niche dynamics, diets, and other life properties of wildlife.

Studies of behavioral data of wildlife species and its relation to the conservation and management of wildlife have become common. Experts in the field have prescribed standard methods for sampling behavioral data as well as additional methods to measure and analyses the same. So, for the study of behavioral ecology, dynamic modeling has assumed increasing importance.

Studies on optimal foraging by wildlife was done by Belovsky. In recent studies on foraging Behavior of wildlife, analysis of trade-offs between obtaining food and risk of losing life to predators is involved. Failure to consider both the critical constraints on Behavior may lead to erroneous conclusions about habitat requirements of wildlife.

Methods have also been evolved to study the effects of social Behavior on ecosystem processes. Models are available to estimate the diets of wildlife species and their effects on ecosystems.

Although a lot of work has already been done in this field, but much more research in this field is still required, including the development of new statistical methods for understanding and managing wildlife populations.

1.6. WILDLIFE AND NUTRITION

Over a period of time, the human population has grown multifold. Therefore, in order to accommodate the growing population and to meet their needs of food and shelter, human settlements have expanded to the extent of drastically modifying the global landscape.

Growth of the human population combined with their ambitions and growing needs for development has led to the transformation of the earth's landscapes and waterscapes. Human activities like urbanization, agriculture, resource extraction, industrialization, energy production and military activities have severely affected and strained the geological resources, forestry, rivers, oceans etc.

As a result of human induced changes over the years, the environment in which the wild animals lived has been changed, including the abundance of food items and their quality, which has adverse implications on the health of animals, their reproduction and survival. The accumulated effect of human induced changes on earth's ecosystems has led to the point, where human activities are being considered to have a dominant influence on the climate and environment.

Recently, awareness towards the need for wildlife conservation and management has led the growth of two fields of nutrition, namely Nutrition ecology and Nutrition physiology.

Nutrition Ecology is the field of study of relationships between diet, digestive physiology and feeding Behavior. On the other hand, ***Nutrition Physiology*** is the study with a focus on the relationships pertaining to intake and assimilation of food items, that is how choices of food and digestion are affected by the abundance and composition of foods in a given environment.

These fields have generated many insights into understanding the mechanisms that connect ecological patterns and processes to animals' appearance, development, and Behavior. Apart from this, the fields of nutrition ecology and nutrition physiology also help in understanding the nutrition induced constraints on locomotion, activity patterns, demography, and population dynamics.

1.6.1. Basic Food Requirement for Wildlife

Like humans, food is the basic need of wildlife, from which they get energy. For proper growth, metabolic activities and breeding, plants, and animals need the energy and nutrition that is derived from the food they consume.

The green plants are autotrophs, that is, they manufacture their food on their own, through the process of photosynthesis. The animals are heterotrophs as they do have the capability or mechanism to manufacture their food on their own and have to depend upon other sources.

Certain animals, which are dependent on plants and consume them to meet their food and energy needs are called herbivores or herbivorous animals. In turn, the animals which are dependent on herbivorous animals and eat them to meet their requirements of food and energy are called carnivores or carnivorous animals. Thus, a food chain is formed in this manner and ecological balance is maintained.

Although all the animals have requirements of nutrition, different species have different food requirements. These food habits are adopted as per the habitat or environment in which they live. A particular food may be edible for a certain species, but non-edible for others.

Therefore, the quality and quantity of food available in a particular habitat have a direct effect on species, numbers, and status of wild animals. The assessment of quality, quantity, characteristics, and status of vegetation available in an environment is an important and basic aspect of wildlife management and conservation

The necessity of food for wild animals is there to meet both their physiological as well as psychological needs.

1. ***Physiological Needs:*** Animals meet their requirement of nutrients in different ways. Also, their requirement may be different in different stages. Certain animals meet their nutritional requirements by directly consuming the plants. But in some animals, the nutrients are synthesized by the presence of certain

bacteria in their digestive tracts and then made available in useful forms of nutritive elements for the body.

In certain stages like during pregnancy and breeding periods, females require more food or some special food to sustain the proper growth of the offspring.

2. ***Psychological Needs:*** Digestive system of a species adapts itself as per the food habits of that particular species. As an example, herbivorous animals develop their eating instincts towards plants or grass, whereas carnivorous animals have the same towards herbivorous animals.

In order to sustain themselves through various seasons, animals consume different types of foods as available and in this way, their food choice is decided.

Thus, the food choice of animals can be classified as follows:

a. ***Preferred Food***: The favorite and most liked food by a species is called their preferred food of that species. For example, bananas are the preferred food of monkeys.

b. ***Staple food***: Staple food is eaten routinely and, in such quantities, that it constitutes a dominant portion of a standard diet for a species, mostly meeting its energy needs and generally forming a significant proportion of the intake of other nutrients as well.

c. ***Emergency Food***: In the event of non-availability or scarcity of staple food, species depend upon certain foods, which may not be to their liking or of less nutritive value. Such foods are called emergency foods and cannot sustain a species for a long period as it may lead to malnutrition.

d. ***Stuffing Food***: Stuffing foods neither fulfill the nutritional nor energy requirements of an animal species. They are consumed by animals to stuff their stomachs when there is no available alternative.

e. ***Miscellaneous Food***: At times, certain animals eat the food that is not part of their usual food habits. Grass eaten by tiger or bark eaten by sambhar or cheetal can be cited as examples of such foods.

Following are the factors that influence the animal specifies for their food selection:

1. ***Availability:*** The food selection by a species is dependent upon the factors that in which season of the year what type of food is available in a habitat and in which quantity, and how much effort is required to be put in by that species to acquire that food.

Usually, the food which is easily available in sufficient quantities in a particular season, is consumed by that species. It can be inferred that abundance of food is directly linked to its consumption.

2. ***Palatability***: For an animal, palatability of food is related to its taste, digestibility, and easy availability.
3. ***Physiology***: Physiological factors, like requirement to nutrients by a species to sustain itself.
4. ***Habit***: Since evolution animals select the food to eat, as per their habitual instincts and their capability to digest it.

After consideration of all such factors and situations, it can be concluded that the abundance that is, quantity, and composition that is, quality of food is of utmost importance for survival as well as propagation of a species.

1.6.2. Nutritional Requirements

The nutrients required by animals include carbohydrates, lipids, nucleic acids, proteins, minerals, and vitamins. It can be said that the nutritional requirements of animals are quite complex and extensive as compared to the nutritional requirements of plants.

1. ***Carbohydrates:*** Carbohydrates are basic source of energy for all animal species. Animals unlike plants, fulfill their carbohydrates' requirements from external sources. More than half to two-thirds of the total energy measured in calories, that every animal consumes on a daily basis is from carbohydrates.

Carbohydrates breakdown into Glucose during the digestive process, which is the main energy source for an animal. Metabolism of glucose takes place during cellular respiration and a part of the energy is used to synthesize adenosine triphosphate (ATP). Other useful carbohydrates are maltose, lactose, sucrose, and starch.

2. ***Lipids:*** Lipids are substances such as fatty acids, neutral fats, oil or wax that dissolve in alcohol but not in water. They contain carbon, hydrogen, and oxygen but have far less oxygen

proportionally than carbohydrates. They are used to form cellular and organelle membranes that surround nerve fibers and certain hormones.

3. ***Nucleic acids***: Animals obtain nucleic acids from plant and animal tissues, especially from cells that contain nuclei. Nucleic acids are used for the construction of deoxyribonucleic acid (DNA), ribonucleic acid (RNA) and ATP. During digestion process, nucleic acids are broken down into nucleotides that are absorbed into cells.
4. ***Proteins:*** Proteins form the framework of animal body and are major components of muscles, ligaments, and tendons and essential components of cytoplasm, membranes, and organelles. They are also an essential constituent of enzymes. Proteins are composed of about 20 kinds of amino acids. Though many amino acids can be synthesized, requirements of many other types are met through diet.

During digestion process, proteins breakdown into their constituent amino acids which are absorbed into body.

5. ***Minerals and Vitamins:*** Phosphorus, calcium, zinc, potassium, magnesium, and sulfur are the minerals required by animals for their growth and physiological needs. Animals usually obtain these minerals on consuming plants. Vitamins are organic compounds, required in traces by animals for their health. Vitamins are either water soluble or fat soluble.

Water soluble vitamins must be consumed frequently, whereas fat soluble vitamins are stored in liver in fat droplets. Some essential vitamins for animal health are Vitamin A for good vision, Vitamin B for substances used in cellular respiration and Vitamin D to assist in calcium absorption in the body.

Animals obtain the required nutrients in different ways and in a vast variety of feeding patterns. For example, Sponges feed on small particles of food that are absorbed in their pores. Sea cucumbers obtain their food by waving their tentacles and trapping food on their sticky surfaces. Mollusks feed by filtering material through mucus in their gills.

Animals that have organs for seizing, chewing, and eating food, feed on food masses. For example, herbivores are animals that feed on plants, while carnivores survive by feeding on other animals. Humans are the examples of omnivores that consume both plants are animals.

6. ***Water:*** Water is an essential requirement for all life forms. It is necessary for digestion, metabolism, cooling, lubrication, and other life processes. Following are the sources of obtaining water for the wild animals:

a. Free water e.g., water from lakes, streams, rivers or vegetation;

b. Water from food consumed; and

c. Metabolism water that is produced during the breakdown of proteins, carbohydrates, and fats.

1.6.3. Body Composition

Knowledge of body composition of wildlife species is of importance due to several reasons:

a) Fat content can be used as an indicator for conditions of animals;

b) Body composition for prey species of animals is a vital factor in order to study wildlife food chains, for the determination of amount of energy;

c) Protein availability in higher trophic level; and

d) Game species of animals are often used for consumption by humans; therefore, knowledge of their body composition is important in the evaluation of the contribution of limiting to human nutrition.

The tissues of animals are composed of water, protein, fat, minerals, and carbohydrates. In a matured animal:

- Water content available varies from 45–60% (depending upon fat content);
- Carbohydrate content is less than 1%;
- Mineral content varies from 2–5%; and
- Protein content ranges between 15–20%.
- The fat content influences the relative percentage of other chemical constituents. This implies that the higher the fat content, the lower is the percentage of other constituents.

Body fat serves a very important function, as a mechanism for energy storage. This is vital for some species for their survival during winters, when the nutritive quality of available food is bad. Body fat deposit plays a very vital role that migrates from one biome to another, to avoid harsh climatic conditions.

For species that hibernate during unfavorable seasons, fat content is a key factor for dormancy during the hibernation period. Hibernating species suffer considerable weight loss during dormancy. Certain species like deer utilizes body fat deposit as an insulator and to supply energy during winters.

Role of fat levels for individual species and community relationships is one of the key research areas in the field of wildlife nutrition.

1.7. WILDLIFE AND ADAPTATION

The environment in which the animals live is a very important factor for their existence. An animal's environment comprises of different elements like climatic conditions, the type of vegetation and also animal species, that exist in the said climatic conditions.

Climate is the most important factor that has significant influence on the food chain of wildlife. Whether the conditions are hot, cold, dry or wet, will influence all wildlife species that live in that particular location.

The type of vegetation that grows in the given climatic conditions, especially the food plants, their variety, abundance, and composition is a key factor for the existence and survival of types of animals in a given environment. If the flood plants available are in good quantity, variety of animal life will be found in that environment.

For example, if in a particular location, predators are around, prey animals have to learn to defend themselves in various ways. The prey animals, which cannot defend themselves, become an easy food for predators. Similarly, the predators have to learn how to hunt the prey that forms their staple diet.

Therefore, adaptation in the context of wild animals is something like a special skill that makes it possible for them to live in a particular place or in a particular way. The adaptation may be physical, that is the way in which the body will work, or the shape or size of the animal's body itself. The adaptation may also be behavioral that is the way an animal will behave in a particular situation. Each and every adaptation comes from the process of evolution.

The animals that cannot adapt to environmental changes, perish. The animals that adapt to environmental changes survive and breed in changed environment. The off springs of such animals are adapted to the new environment and the life cycle continues.

Therefore, the skill of adaptation makes it possible for a great variety of wildlife to live and thrive on the earth. Most of the animals adapt themselves to the natural conditions. Wild animals must have the right kind of habitat, where they can find food and space they need for their existence.

1.8. WILDLIFE SHELTER

The place or area in which a wildlife species lives, or a place/area that gives protection to a wildlife species and caters to their other biological needs, is known as shelter. Shelter may be in the form of a cover or a space.

Covers are those vegetation or plants where animals take shelter to protect themselves from various climatic factors. Covers provide safety to animals and are used by these for shade and taking rest.

The covers can be classified as:

a) **Escape Covers** are covers where animals hide to protect themselves, after escaping from predators or hunters. Their shape and size depending upon the size of the animals.

b) **Nesting or Breeding Covers** are covers which birds use to hide themselves for laying and nursing the eggs and breeding. Large animals also need such covers for breeding. Its shape and size also depend upon the size and requirements of birds and animals.

c) **Shelter Covers** are covers that protect the animals from adverse seasons and adverse climatic conditions. It may be bushes, trees or caves.

d) **Roosting Covers** are covers where animals take rest after taking food and use as perch. These may be bushes, trees or caves.

1.8.1. Space for Wildlife

Animals need sufficient space for living It is an important part of a habitat and without its survival of animals becomes difficult. The space may be multi-dimensional, horizontal or vertical as per the requirement a particular animal species. In the horizontal dimension, sufficient space is required as per the need of the animals like home range, territory, etc.

Certain wild species need space in vertical dimension in any field. In vertical dimensions, wild species reside in six or seven levels. For example, vulture fly at the peak height, then eagles, then crows and some types of birds at lower levels in sequence, who live on trees. Below them, some small birds, squirrels, monkeys live on branches of trees. Some animals make their dwelling in hollows of tree trunk.

Similarly, on the ground, the terrestrial animals like mammal's roam followed by rats, reptiles, and insects that dwell underground in the soil.

Due to the non-availability of sufficient space, the animals are stressed leading to the possibility of a crowd of animals in a habitat. This leads to possibilities of casualty, spread of disease amongst the animals, adverse effects on breeding, improper supplies of food, water, etc.

1.9. WILDLIFE ADAPTATION AND CLIMATE CHANGE

As per a study conducted in recent years, the change of climate is already showing its adverse effects and significantly damaging living organisms and ecosystems existing on land and water. These climatic changes have led to rising temperatures, changes in precipitation patterns and weather becoming more extreme and less predictable and due to this, animals are shifting their range and altering the timings of key stages of their life cycle. In addition, these are also showing adverse effects on their sex ratios, tolerance to heat and on their bodies. These changes may support some species to adapt, but others could perish.

1.9.1. Move, Adapt, or Die

Various animal species can respond to changes in climate in three ways only: Either they can move to a new suitable location or adapt to the climate changes to survive, or they die or perish.

Many animals may be moving to higher elevations or latitudes to escape the rising temperatures, but maybe the climatic changes are taking place too fast and quickly for most of the species to beat it. Moving to a new place has its own problems and may not be an easy solution.

Entering an unfamiliar new territory could mean interacting with unfamiliar species or conditions or facing more completion for food. Some species like American Pika, a hamster, which are living at higher altitudes in cool and moist conditions, may soon not find a place to go to higher elevations due to rising temperatures.

Also, in some cases, animals which are moving to places with cooler climatic conditions may be restricted by highways or other man-made structures.

However, certain species like Monarch butterflies are not able to outrun the impacts of rising temperatures. These are now threatened by extinction and over the last 20 years, their numbers have declined by 95%.

To explain in brief, monarch butterflies fly southwards from Canada towards Mexico to spend their winters, taking cue from day's length and temperature. However, of late, their southern migration is delayed by more than one and a half months as warmer than normal temperatures, do not cue them to fly south. And with the onset of cooler temperatures in Mexico in spring, they return towards north to lay their eggs. As temperatures increase, their migration could fall out of sync with the bloom time of the nectar producing plants, they depend upon for food.

Due to the rise of temperature in the Arctic region resulting in the melting of sea ice, polar bears are losing, not only their food source that is seals, for which they hunted from sea ice, but also the place to rest and breed.

Another species facing threat to their survival from global warming are Puffins, that live in the Gulf of Maine. Their staple food is white hake and herring. But as oceans are getting warm, these fish are moving farther north. As a result, the puffins are trying to feed their young on butterfish for survival. But as young puffins are not able to swallow butterfish that are larger, they are starving to death.

1.9.2. Species That Adapt

There are certain species that appear to be adapting to the changing climatic conditions and rising global temperatures. Due to rising temperatures, the spring season has been advanced that is, it arrives earlier and thus the insects emerge earlier. To match with the insect availability, some migrating birds lay their eggs earlier, so that the newborn birds can have the food.

A study on breeding patterns of female butterflies in Southern Australia indicates that over a period of last 65 years, the time period for their emerging from their cocoons has advanced by 1.6 days per 10 years whereas the temperature rises in the area had been 0.14°C per 10 years.

Coral reefs, which are colonies of tiny soft bodied organisms called coral polyps get their wild hues from billions of zooxanthellae algae, which they host. The rise in ocean temperature or due to pollution, these get heated up and have experienced extensive bleaching, as the coral polyps release zooxanthellae algae. Coral bleaching can extensively damage the entire colony.

However, scientists studying coral reefs around American Samoa have found out that many corals in pools of warm water have not bleached. Lab experiments were carried out on these corals by further exposing them to still higher temperatures and it was observed that only 20% these expelled the algae.

Similarly, when corals from cooler pools were exposed to high temperatures, 55% of them expelled the algae. However, when coral pools were moved to pools of warm water, only 32.5% of these expelled the algae, as their tolerance towards heat improved. They adapted to their new environment without any genetic change.

The above research done on coral has demonstrated the difference between adaptation over the course of several generations, through natural selection that is, evolution, and adaptation through phenotypic plasticity.

Phenotypic plasticity, in simple terms can be defined as the ability of an organism to change its developmental, physical, and behavioral features during its life span, in response to the changes in environmental conditions.

Generally, phenotypic plasticity is more important for immobile organisms like plants, rather than mobile organisms, as mobile organisms can often move away from an unfavorable environment.

Thus, the corals living in the warm water pools had evolved over many generations as the natural selection favored the survival of the most heat tolerant corals and enabled them to reproduce. But the corals from cool water pools, when they were exposed to hotter conditions were also able to adapt to warmer conditions for survival, as they had phenotypic plasticity.

When some animal or plant specifies come across the effects of changes in climatic conditions in the environment they inhabit, they respond by changing their Behavior and moving to the zones that are cooler, modifying their physical bodies in a way, to deal better with the hot conditions, or changing the timings of their certain activities to match with the changes in the seasons. Such "plastic" changes take place as some genes can produce multiple effects when exposed to different environmental conditions.

- **Epigenetics** is the study of heritable phenotype that do not involve alterations in the DNA sequence. In simple terms, it is the study of how the environmental factors affect the genes, cause them to switch on and off, and bring about phenotypic plasticity mainly through producing certain organic compounds that attach DNA or modifying the proteins that are wound around.

This determines when and how a gene is expressed, but it does not alter the DNA sequence itself. In some cases, such changes are passed along to the next generations. But if environmental stresses are eliminated, the epigenetic changes can be reversed.

Still it is not known that all the species have the capability and capacity of epigenetic responses. But for the species that have this capacity, epigenetic changes could make more time available to evolve genetic adaptations to changing environmental conditions.

In the long term, phenotypic plasticity could become an evolutionary adaptation, if the individual species with genetic capacity for phenotypic plasticity are better suited for survival and higher reproduction in the changed environmental conditions.

Some experts are of the opinion that like any other trait, phenotypic plasticity can undergo a natural selection. Thus, when it is beneficial to have a plastic response to the environment, this can be favored by natural selection. Some traits, like behaviors, may be more likely to be plastic than others.

Species whose reproduction is not frequent and that take a long time to mature, epigenetics may give more flexibility to such species to enable them to adapt to fast changing conditions. It is seen that species which have shorter life span reproduce more frequently, and the rapid succession of generations helps such species to evolve genetic adaptations through natural selection much more quickly.

1.9.3. Epigenetic *Changes: A Few Examples*

Following two studies done by scientists illustrate the examples of epigenetic changes in terrestrial and aquatic species:

a) Scientists conducted a study on populations of winter skate fish from two locations that is, the Atlantic coast and Southern Gulf of St. Lawrence. The skates from gulf are smaller in body size than those living on Atlantic coast. The waters at gulf are warmer by 10 degrees Celsius as compared to the Atlantic coast.

It was found that skates living in gulf adapted to warmer temperatures by reducing their body size by 45% as compared to other populations, to meet their oxygen requirement, as oxygen content is less in warmer oceans.

Scientists detected more than 3000 changes in gene expression that explained changes in body size and some physiological traits. However, despite these epigenetic changes, the DNA of both populations of winter skates were found to be identical.

b) Another study was conducted on guinea pigs found in South America. These guinea pigs normally mated at a temperature of 5 degrees Celsius. After keeping the male guinea pigs at a temperature of 30-degree Celsius, epigenetic changes in 10 genes related to body temperature was detected. The offspring's of these guinea pigs showed epigenetic changes, but these were different from their earlier generation.

It was inferred that while the fathers showed epigenetic changes in response to higher temperatures, but the changes passed on to their offspring's were towards preparing them for living in warmer temperatures.

1.9.4. Examples When Phenotypic Plasticity Is Not Protective

1. **Maladaptation:** Meadow voles (a species of field mouse found in North America) born in autumn of have a thicker coat as compared to those born in the spring season. This is due to the environmental cues given by female voles to unborn offspring's in their womb through hormones.

These predictive adaptive responses, as believed to be governed by epigenetics, guide the animal's metabolism and physiology that enables it to adapt itself to the environment, in which it will be born. Though this adaptation is suitable for them to be born in normal winters, it could end up being maladapted, if winters become warmer.

2. **Limit Adaptive Evolution:** Phenotypic plasticity in seasonal habitats may give birth to species that will live and thrive in that particular environment, but these are more vulnerable to climate changes. For example, a Malawian butterfly has high reproduction rate and a shorter life span, if it is born in warm, wet time of a year. But if born during cool, dry season it has an inactive long life with delayed reproduction.

Scientists have found that while a butterfly has a variety of gene expressions, very little gene variation has been observed for this plasticity. The butterflies are adapted to very consistent, specific, and predictable environment cues. Natural selection further tuned these carefully tuned responses. Over a period of time, this natural selection eliminated the genetic variation, that would have allowed more plasticity.

It is also seen that the tropical species have lesser diversity in their genes to deal with changing conditions as the tropical regions have held very little variation in their climatic conditions over a thousand of years. Therefore, it is believed that the species in regions having consistent climate throughout the year, find it more difficult to adapt to variations in climatic conditions.

1.9.5. Evolution to Rescue

The research is also being done on global patterns of coat color changes in eight species of hares, weasels, and foxes. It is found that it is more common at higher altitudes to find individuals that turn white in the winters. But with less snowfall, the mismatch of their white coats with surroundings has led to reduction in their range.

Dr. L. Scott Mills, Professor of wildlife biology at the University of Montana, U.S.A. observes that these animals either having brown or white coats in winter neither have a strong genetic component, nor the trait of change of coat color has behavioral plasticity, they do not seem to have the capacity to behave so as to reduce mismatch or reduce their ranges, being killed by the mismatch.

With decrease in snowfall such mismatches of color will become more evident, hence in order to survive, they have to evolve. The research identified some populations of these animals, in which some individuals that turn white and others that stayed brown in winter.

As these groups have genetic variability, they have a good possibility to adapt, since the process of evolution is faster in cases where there is enough variation in a population for natural selection to act upon.

But the phenomena of both evolutionary change and phenotypic plasticity are more likely to occur in those animals who have large populations and those who have a link with other populations.

It is because a large diverse group will have more individuals with genes that allow phenotypic plasticity, which can ultimately be favored by natural selection. Apart from this, the species which can live in an environment with the varieties of conditions will usually have more variations in their traits that can be inherited.

1.10. CONCLUSION

If we do not reduce our carbon emissions and allow the global temperatures to continue to rise, the climatic change that takes place as a result of temperature rise could pose a very serious threat to the survival of plants and animals species, even in the world's most biodiverse areas.

A number of methods backed by multiple public domain software have been deployed in the field of wildlife ecology to study the critical elements of wildlife population, especially the rare and endangered species. Also, a lot of research work is being done by the scientists on the nutritional needs of wildlife and in the field of genetic diversity needed for phenotypic plasticity and adaptive evolution. But a lot more needs to be done.

The variety of life on earth and the proper functioning of its ecosystems is important for all human beings. If other species are not able to adapt changes in climatic conditions, the consequences could pose dire threat to the humans. Hence the strategies need to be developed, that will help the wildlife to adapt to climate changes for their survival.

This means identification and protection of zones where specifies exhibit genetic variability and preserving land-based and marine-based ecosystems. This implies purposefully increasing the connectivity between natural areas and also identifying and making available areas of land to facilitate the migration of animals. These actions will help the species to move to the cooler areas.

This will also allow for larger and more connected populations, that can promote genetic diversity needed for phenotypic plasticity and adaptive evolution.

REFERENCES

1. Bowyer, R., Stewart, K., Monteith, K., & Long, R., (2013). Wildlife ecology is based in part on the article "wildlife ecology" by R. Terry Bowyer, which appeared in the encyclopedia of environmetrics. *Encyclopedia of Environmetrics,* [online] Available at: https://www.researchgate.net/publication/319264629_Wildlife_ecology (accessed on 21 August 2020).
2. Boyce, M., (2012). *Wildlife Ecology*. Oxford bibliographies online datasets, [online] Available at: https://www.oxfordbibliographies.com/view/document/obo-9780199830060/obo-9780199830060-0002.xml (accessed on 21 August 2020).
3. Cho, R., (2018). *What Helps Animals Adapt (Or Not) To Climate Change?* [online] State of the Planet. Available at: https://blogs.ei.columbia.edu/2018/03/30/helps-animals-adapt-not-climate-change/ (accessed on 21 August 2020).
4. Cliffsnotes.com. (n.d.). *Nutrition in Animals.* [online] Available at: https://www.cliffsnotes.com/study-guides/biology/biology/nutrition-and-digestion/nutrition-in-animals (accessed on 21 August 2020).
5. Definitions.net. (n.d.). *What Does Wildlife Mean?* [online] Available at: https://www.definitions.net/definition/wildlife (accessed on 21 August 2020).
6. Dotson, J., (2019). *Biome: Definition, Types, Characteristics and Examples.* [online] Sciencing. Available at: https://sciencing.com/biome-definition-types-characteristics-examples-13719223.html (accessed on 21 August 2020).
7. Khan Academy, (n.d.). *What is Ecology? (Article) | Ecology | Khan Academy.* [online] Available at: https://www.khanacademy.org/science/biology/ecology/intro-to-ecology/a/what-is-ecology (accessed on 21 August 2020).
8. Meenakshi, G., (n.d.). *Essay on Wildlife: Top 6 Essays | Natural Resources | Geography*. [online] Geography Notes. Available at: https://www.geographynotes.com/essay/wildlife/essay-on-wildlife-top-6-essays-natural-resources-geography/3940 (accessed on 21 August 2020).
9. National Geographic Society, (n.d.). *Ecosystem*. [online] Available at: https://www.nationalgeographic.org/encyclopedia/ecosystem/ (accessed on 21 August 2020).

10. Uen.org. (n.d.). *Animal Adaptations-UEN.* [online] Available at: https://www.uen.org/themepark/habitat/animal.shtml (accessed on 21 August 2020).
11. *Wildlife Nutrition,* [ebook] Available at: http://unaab.edu.ng/wp-content/uploads/2009/12/465_WRM%20506%20Note.pdf (accessed on 21 August 2020).

2

Population Fluctuations in Wildlife

CONTENTS

Population fluctuation in wildlife is one of the most common problems and is prevalent since the past few decades. In the chapter population fluctuations in wildlife, the various factors that contributed to population fluctuation is discussed.

Some of the factors such as limiting factors, human factors and natural factors are being discussed. life tables and the rate of population growth wildlife population parameters population fluctuation and population cycles the basic components of population change wildlife population management historical background and scientific foundations the impact of population growth on wildlife population trends in protected areas three main factors that control populations.

2.1. INTRODUCTION

Population, in human biology, refers to the total number of inhabitants living in a particular area (such as a country or the world) and persistently being reformed by expansions (such as births and immigrations) and losses (such as deaths and emigrations). Just like any other biological population, the range of a human population is restricted by the total food supply, the outcome of problems or dangers such as disease, and other environmental factors.

There are various other factors that affect the overall human populations such as social customs leading reproduction and technological advancement in a given period, primarily in public health and medicine domain, that have lessened mortality rate and extended the life span.

There are some aspects of human societies that are as fundamental as the size, composition, and rate of variation of their populations. All these factors are responsible for affecting health, economic prosperity, family organization, education opportunities, crime rates, culture, language—indeed, virtually every aspect of human society is touched upon by population trends.

The study which is concerned with human populations is known by the name demography—a discipline with intellectual origins dated back to the 18th century, when it was initially acknowledged that human mortality could be estimated by using the statistical techniques.

Demography casts a multidisciplinary net, drawing insights from sociology, economics, medicine, statistics, anthropology, biology, and history. In addition, it was also found that its chronological sweep is lengthy: inadequate demographic support for many centuries into the past, and trustworthy data for the past numerous decades are available for many regions.

The recent knowledge about demography makes it feasible to project (with caution) variation in population several decades into the future.

2.2. POPULATION

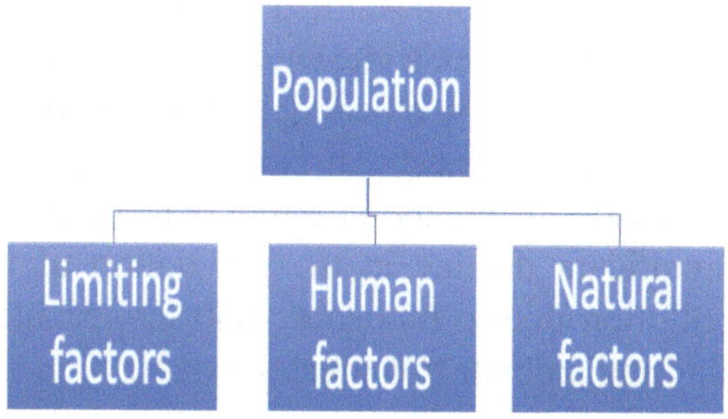

Figure 2.1. *Factor affecting wildlife population.*

Population refers to the total number of individuals of a certain species living in an area. The population of plants and animals depends on the number of factors for survival. Limiting factors such as the availability of water, food, and shelter facility can impact the overall population of the organism.

2.2.1. Limiting Factors

In nature, populations of plants and animals are associated with each other like a puzzle. The pieces of the animal and plant puzzle vary on each other to form a complete picture. If one entity is facing any problem, then the whole group has to suffer.

In the natural world, limiting factors such as the availability of water, food, shelter facility, and space can alter the plant and animal populations to a greater extent. Other limiting factors, such as the competition for valuable resources, disease, and predation rate can also impact the overall populations. If there is a change in any of the limiting factors, then the population of both plants and animals will change. However, the rate of change will depend on the number of internal and external factors.

Some changes may result in increasing the population. For instance, if the number of plants in a given area is more, then the population of animals will likely increase as they eat those plants. If the population of one animal increases, the population of animals that eat that animal might also increase.

2.2.2. Human Factors

Humans can impact both the plant as well as animal populations. When humans build houses or buildings in open areas such as forests, then they cut down trees in large amounts, which resulted in changing down the habitat of both plants and animals. However, some animals such as raccoon and skunk, have the potential to adopt, while others do not, resulted in impacting their population. The common loon nests on land near large lakes. Because of extensive human activities, some loon nesting places have been eradicated, resulted in the significant decline in the population of the loon. It was also noted that pollution due to human activity contributes to impact the population of plants and animals to a greater extent.

2.2.3. Natural Factors

It is generally seen that predator/prey relationships play a huge role in the population of animals. If there is any alteration in the balance between predator and prey, then populations change can be clearly observable. The population of white-tailed deer has grown substantially in some areas because there are no natural predators. Wolves and mountain lions are the natural predators of the white-tailed deers. Their population-has declined because of overhunting and habitat loss. All these factors are outside the scope of humans and no one can control them.

2.3. LIFE TABLES AND THE RATE OF POPULATION GROWTH

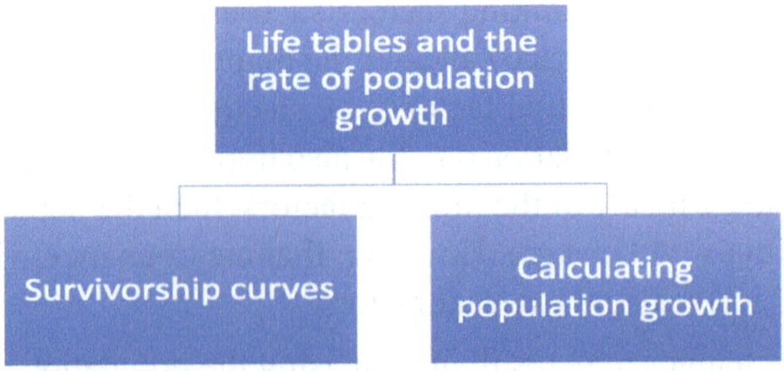

Figure 2.2. *Life tables and the rate of population growth.*

Differences in the strategies of life history, which involves the allocation of the organism of its time and resources to the process of reproduction and care of offspring, greatly have an impact on the population dynamics. As it is stated above, populations in which the person goes through the process of reproduction at an early age have the potential to grow at a much faster pace as compared to the populations in which individuals reproduce at the later stage.

The impact of the age of first reproduction on the growth of the population can be seen clearly in the life tables for a specific species. It has been observed that the life tables were developed originally by insurance companies to provide a way of determining how long an individual of a particular age could be likely to live their life.

It has to be noted that they are not just used by demographers of human populations but they are also used by animal, plant, and microbial ecologists. They are used in order to make projections regarding the life expectancies of nonhuman populations. In addition to it, it is also used to make projections about the effects of variation on demography as well as on the population growth.

The total number of people in a closed population is governed by the rates of birth (that is known as natality), reproduction, growth, and rate of death (that is known as mortality). A closed population is a population in which neither immigration nor emigration takes place. Life tables are designed to calculate how these rates have impact on the entire growth rate of a population.

2.3.1. Survivorship Curves

Life tables follow the fate of a group of people who are born within a similar population and also, in the same year. Of this group of people, or cohort, just a certain number of people will reach each age, and also, there is an age above that no people ever live.

Formulating the number of those individuals of the group that are still alive at each age gives rise to a survivorship curve for the population. Usually, survivorship curves are displayed on a semi logarithmic scale. These curves are not displayed on an arithmetic scale.

It has to be noted that there are three general kinds of survivorship curves. Species like humans and other large mammals have fewer offspring but these species devote much time as well as their energy in caring for their young or offspring (K-selected species).

This curve is relatively flat. It reflects low juvenile mortality, with the majority of the people living to old age. A constant possibility of dying at any age or time of life that is shown by the Type II survivorship curve is evident as a straight line. This straight line has a constant slope that reduces over time toward zero.

In addition to it, apart from the species that are discussed above, certain lizards, rodents, and perching bird also display this kind of survivorship curve. There are some of the species that produce many offspring but they invest little care and less time for them (r-selected species). In such species, mortality is greatest among all the youngest individuals.

The survivorship curve, which is used to indicate this life history, has a very steep path in the initial stages. This reflects that there is a very high mortality rate among the young. On the contrary, this curve but flattens when used for indicating the individuals who reach maturity. This means the individuals reaching maturity survive for a relatively long time.

This kind of life pattern can be observed in many families of insects or shellfish. Most of the populations have survivorship patterns that are more complicated than, or fall in between, these three idealized curves.

For instance, passerine birds (these are the perching birds like finches) generally suffer high mortality during the first year of their life and a lower. In the following years, there is more constant rate of death.

2.3.2. Calculating Population Growth

In order to study the population growth, life tables are also used. The average number of offspring that are left by a female at each age together with the percentage of people who are living to each age can be used in order to calculate the rate at which there is a change in the size of the population with the time.

Demographers and population ecologists make use of these rates. These rates are used in order to estimate the growth of the population and to calculate the impacts of conservation efforts on all the species that are in the fear of extinction.

2.4. WILDLIFE POPULATION PARAMETERS

Figure 2.3. *Various parameters to keep a check on wildlife population.*

Source: Image by Flickr.

Among all the ways, one of the easiest as well as most convenient ways of assessing or predicting the growth of population, decline of the population, or stability is to calculate the percentage of young animals to old in a population. This measure is known as the population's age ratio.

Then, this information can be shown graphically in an age pyramid. Age ratios are generally used in order to compare all the variations in a single population between years or within a similar year for different populations. This method has been used by wildlife agencies. They have used this method to calculate the increases in white-tailed deer in the previous times.

Many methods can be used to reflect the information. Computer models as well as simulations can also be used in order to reflect this information more accurately. Recent information has suggested that age ratios may not reflect the real or the actual variations in the populations.

Though, age pyramids can permit a private property-owner to take a quick look at the number of young animals against the old animals that are present in the population and dispel various myths that are related to the stockpiling of wildlife.

Even though there can be exceptions, but there can be decline in the population of wildlife if they are top heavy or support a much larger number of adults than yearlings. This decline in the population of wildlife, over time, can be attributed to few numbers of new individuals that are moving into the reproductive section of the population.

On the contrary, it has to be kept in mind that a young population will have a greater percentage of yearlings. There will be increase in the population as the maturation of the yearling class expands the reproductive section of the population.

Often, the observation from evaluating the population age structure are applied to long-lived species (white-tailed deer) and it usually needs extensive efforts in order to get the whole age distribution for the population of animals.

With comparatively short-lived species (e.g., quail, woodcock, muskrats, cottontail rabbits, mourning dove), the information is gathered more easily but this information will be beneficial only into the next year. This is because of the high rate of death and the fast turnover of the populations.

For small game, the age pyramids show that a large proportion of the population never makes it to age 3. This informs the owner of the land and manager that they cannot stockpile game (by not shooting or hunting in the hope of having a larger breeding population in spring) until and unless there is enough habitat. There is a requirement of sufficient habitat in order to support the animals.

In addition to it, it also provides the other important lesson: it is tough to increase the population of small game bird by stocking pen-raised birds (majority of these birds will generally die over the first winter) until and unless there is the availability of habitat in order to support a larger population.

The birds may be present if the habitat is present. Sex ratios are the other aspect of populations that is examined by the manager of wildlife carefully. This is because a disturbance in the percentage of males to females can impact the reproductive success of a population.

Often, the sex ratios are expressed as the percentage of males that are present in the particular population or the total number of males per female in the population. The information about sex ratio is used by wildlife agencies at the time of handling big game herds.

Wildlife agencies manipulate the sex ratios of bucks to does that are eradicated from the population every single year in an attempt to make the maximum number of creatures that can be harvested. Also, there should be the maintenance of an adequate number of bucks in the population in order to assure a complete reproductive effort by does the next year.

In case of polygynous species, such as the white-tailed deer (one male can breed with various females), sex ratios are skewed or these ratios are shifted in order to favor females that are present in a harvested population. On the other hand, in case of monogamous species, such as Canada geese or coyotes (one male breeds with just a single female), a balanced sex ratio (that is 50:50) is needed for maximum production of offspring.

For instance, in case of Canada geese, if the sex ratio is shifted to 40 males per 60 females, the whole population will display just 66% production. It has been observed that a sex-specific hunting season could devastate a population. This is the case of a monogamous species.

Though, even a wildlife population that is un harvest do not usually maintain an even ratio of sex. Even though the information or the data do not always support the following generalizations, many studies have shown that most mammals give birth to somewhat more males than to the females.

On the other hand, for birds the pattern is totally reversed, the mammals give birth to more females than males. These trends tend to move toward a more even or 1:1 ratio. This happens at the later stages of the life.

2.4.1. Mortality Factors

The events of birth and death are common to all living creatures. It has been observed that animals in the wild die because of the old age. Instead, these animals succumb to one or several factors that have impact on the members of their specific species. Mortality is described as inherent loss of a person from a population through death.

Mortality is hard to evaluate. This is because carcasses are difficult to locate. The fate of animals that that separate or shift somewhere from a population can rarely be determined. Therefore, it is more practical to calculate survivorship, or the total numbers of creatures that are still alive, as these individuals can be situated and accounted for.

The species in the wildlife that are having a high reproductive potential, like cottontail rabbits, seem to have low survivorship. Also, the high mortality at the younger age of their lives. Thus, offsetting their high reproductive outputs.

On the other hand, animals that are having a lower reproductive potential, like white-tailed deer, have higher survivorship of young. Thus, compensating for the smaller litter sizes. Often the survivorship values or rates are got from field data that have the information about different age classes in a population.

With the help of these values, there can be the development of survivorship curves for a particular population. These are used in order to evaluate the potential of the population for increase or decline in a specific habitat.

There are various mortality factors (these includes, predation, accidents, harvest, disease, exposure to severe weather, starvation or malnutrition,) that can be the proximate reason for how an animal meets its fate. While the proximate reason may be predation or exposure to severe weather, the ultimate reason may be related to the absence of enough cover in order to escape predators or the impacts of adverse weather.

The population in the wildlife will be subjected to many limiting aspects with time that may produce an overall decrease in the size of the population. Some factors have a significant influence in one year may have less influence in the following year.

This is because weather patterns, habitat conditions, and populations of predators and prey are changing constantly. Despite all these inconsistencies, it has been observed that in most of the cases, the overall decrease in the size of the population from mortality factors across years continue to be the same.

As a result, the particular reason of death seems to balance or compensate each other. The professionals of Wildlife call this phenomenon compensatory mortality. This is also stated in some other way.

One kind of mortality largely substitutes the other kind of mortality in the population of animals, while the total mortality rate of the population continues to be constant. For instance, bobwhite quail face problems in living in severe weather conditions at the northern limits of their range.

It has been observed that many animals die because of the exposure, and few of the animals will be taken by predators when winters are extremely cold. It has also been observed that at the time of mild winters, many quail live just to fall prey to a host of predators because of limited amounts of escape cover.

The entire impact is that almost the same number of quail are preserved and maintained by the habitat from year to year. The number of creatures that can live in a population, or animals that produce a "surplus." is determined by the habitat to a large extent. Mortality removed this surplus.

Let's take an example. Let's assume that total mortality on the population of a bobwhite quail is 70%; there will be the death of 70% of the entire population in the first year. When the weather is severe in the first year, 20% of the population of bobwhite quail is lost to predators and there is the death of 50% of the population because of the exposure.

70% is the total mortality rate for the population. Then, in the second year, when 60% of the population is removed because of predation, just 10% of the population is lost because of the exposure for a 70% of a combined total mortality rate. The managers of wildlife employ the concept of compensatory mortality at the time of establishing hunting as well as trapping regulations.

In case of compensatory mortality, hunting as well as trapping serve to change the factors of natural mortality that are operating on a population, and it helps in keeping the population density in balance with what the habitat can efficiently support.

In the example of quail, the population was hunted, and there was the removal of 30% of the population. In this scenario, 10% of the population was removed by predators, and 30% of the population of quail died because of the exposure. The total mortality rate did not change. This rate continues to be at 70%.

A harvestable surplus is defined as the part of a wildlife population that is capable of being removed. It is the part of the population that would habitually die because of the other causes.

Data suggested that the species that are having a high reproductive potential can have a larger proportion of the population harvested in any one year as compared to the species that are having a low reproductive potential. The reason behind this is that their higher reproductive outputs will replenish quickly the loss of animals.

Muskrats are considered as a perfect example. When there is the removal of large numbers of muskrats by trapping, the population responds in these ways that are discussed below:

- The breeding season is increased by substantial so there is the production of more litters;

- The number of offspring that are produced per litter increases; and
- Muskrats that are surviving are less susceptible to illness or predation.

Irrespective of the species that is discussed above, an individual must regulate the removal of animals because there is a harvest level in all populations at which the factors of mortality are no longer compensatory but these are additive. This means that there is the addition of one type of mortality to the other sources of mortality.

It has to be noted that additive mortality can be detrimental and it can also lead to decline in the population. Back to the example of quail: it is known that almost 70% of the population will die every single year, whether the population is hunted or not. If the pressure of hunting is heavy and a large ratio of quail is killed, increasing the total mortality above 70%, this would be considered as an additive mortality.

2.5. POPULATION FLUCTUATION

Populations rarely grow smoothly up to the carrying capacity and then remain there. Instead, fluctuations in population numbers, abundance, or density from one-time step to the next are the norm. Population cycles make up a special type of population fluctuation, and the growth curves in population cycles are marked by distinct amplitudes and periods that set them apart from other population fluctuations.

In a few species, such as snowshoe hares (*Lepus americanus*), lemmings, Canadian lynx (*Lynx canadensis*), and Arctic foxes (*Alopex lagopus*), populations show regular cycles of increase and decrease spanning a number of years.

The causes of these fluctuations are still under debate by population ecologists, and no single cause may provide an explanation for every species. Most major hypotheses link regular fluctuations in population size to factors that are dependent on the density of the population, such as the availability of food or the activities of specialized predators, whose numbers track the abundance of their prey through population highs and lows.

2.5.1. Factors Affecting Population Fluctuation

Figure 2.4. Interlinked factors affecting population of wildlife.

Source: Image by Hill Air Force Base.

Population ecologists commonly divide the factors that affect the size of populations into density-dependent and density-independent factors. Density-independent factors, such as weather and climate, exert their influences on population size regardless of the population's density. In contrast, the effects of density-dependent factors intensify as the population increases in size.

For example, some diseases spread faster in populations where individuals live in close proximity with one another than in those whose individuals live farther apart.

Similarly, competition for food and other resources rises with density and affects an increasing proportion of the population. The dynamics of most populations are influenced by both density-dependent and density-independent factors, and the relative effects of the factors vary among populations.

Density-independent factors are known as limiting factors, while density-dependent factors are sometimes called regulating factors because of their potential for maintaining population density within a narrow range of values.

2.5.2. Population Cycles

Because many factors influence population size, erratic variations in number are more common than regular cycles of fluctuation. Some populations undergo unpredictable and dramatic increases in numbers, sometimes temporarily increasing by 10 or 100 times over a few years, only to follow with a similarly rapid crash.

For example, locusts in the arid parts of Africa multiply to such a level that their numbers can blacken the sky overhead; similar surges occurred in North America before the 20th century.

The populations of some forest insects, such as the gypsy moths (*Lymantria dispar*) that were introduced to North America, rise extremely fast. As with species that fluctuate more regularly, the causes behind such sudden population increases are not fully known and are unlikely to have a single explanation that applies to all species.

The size of other populations varies within tighter limits. Some fluctuate close to their carrying capacity; others fluctuate below this level, held in check by various ecological factors, including predators and parasites.

The tremendous expansion of many populations of weeds and pests that have been released into new environments in which their enemies are absent suggests that predators, grazers, and parasites all contribute to maintaining the small sizes of many populations.

To control the explosive proliferation of these species, biological control programs have been instituted. With varying degrees of success, parasites or pathogens inimical to the foreign species have been introduced into the environment.

The European rabbit (*Oryctolagus cuniculus*) was introduced into Australia in the 1800s, and its population grew unchecked, wreaking havoc on agricultural and pasture lands.

The myxoma virus subsequently was released among the rabbit populations and greatly reduced them. Populations of the prickly pear cactus (Opuntia) in Australia and Africa grew unbounded until the moth borer (*Cactoblastis cactorum*) was introduced. However, many other similar attempts at biological control have failed, illustrating the difficulty in pinpointing the factors involved in population regulation.

2.6. THE BASIC COMPONENTS OF POPULATION CHANGE

Figure 2.5. Helping endangered species in national parks.

Source: National Park Service.

At its most basic level, the components of population change are few indeed. A closed population (that is, one in which immigration and emigration do not occur) can change according to the following simple equation:

The population (closed) at the end of an interval equals the population at the beginning of the interval, plus births during the interval, minus deaths during the interval. In other words, only addition by births and reduction by deaths can change a closed population.

Populations of nations, regions, continents, islands, or cities, however, are rarely closed in the same way. If the assumption of a closed population is relaxed, in-, and out-migration can increase and decrease population size in the same way as do births and deaths; thus, the population (open) at the end of an interval equals the population at the beginning of the interval, plus births during the interval, minus deaths, plus in-migrants, minus out-migrants.

Hence the study of demographic change requires knowledge of fertility (births), mortality (deaths), and migration. These, in turn, affect not only population size and growth rates but also the composition of the population in terms of such attributes as sex, age, ethnic or racial composition, and geographic distribution.

2.7. WILDLIFE POPULATION MANAGEMENT

Wildlife population management refers to any strategy that seeks to maintain a target population at a level that can be supported by the ecosystem. This can involve protecting a threatened population from declining further in numbers, or even re-stocking a population.

Conversely, when the numbers of a target population have become too great to be sustained by the food or territory available, then predators can be introduced, or a human-mediated cull can be done.

Culls have also been done when an infectious disease is present in a population; the deliberate killing of the infected animals can help protect other members of the population as well as other species in the same habitat.

Put another way, the management strategies focus on the habitats of the species of concern. Wildlife population management can also use the legal system to prosecute those found responsible for illegally hunting or trapping species.

Figure 2.6. Balancing hill AFB wildlife population with human activity.

Source: Hill Air Force Base.

2.7.1. Historical Background and Scientific Foundations

In the United States, there were few laws or regulations on the hunting of wild animals before 1900. Virtually all animal and bird species were hunted throughout the year.

Although regulations were introduced to control the timing of hunting of species for commercial purposes and to assign quotas on species killed, and, subsequently, similar controls on recreational hunting, various species came under threat from the changing land use in the country.

Territory was lost as land was converted to agricultural, industrial, and urban uses. Furthermore, in agriculture-intensive states, predators such as foxes were actively hunted and killed, allowing their natural prey to increase tremendously in numbers.

By the mid-twentieth century, the need to more actively deal with wildlife populations had been recognized, and the discipline of wildlife biology had been created.

Successful wildlife population management adopts what is known as a landscape perspective. This view takes into account the biological diversity of the particular ecosystem, not just the species of concern.

Although in the case of a species at risk the immediate goal is to stabilize the population, the landscape approach recognizes that by addressing the stability of the entire ecosystem, the threatened species also benefit.

Increased numbers of a particular species can upset the balance of an ecosystem, as the burgeoning population requires more food and, if individuals of the species each require substantial territory, can exceed the available space. The limit of an ecosystem on the numbers of any particular species is known as the carrying capacity. There are different carrying capacities for different species.

Carrying capacity is determined by whatever factor limits the particular population. For wolves, the limiting factor can be territorial size, while for deer the limiting factor can be the available food.

A wildlife population management strategy that can be very effective is hunting. A well-known example is the hunting of deer that takes place in the autumn. Hunters must possess a current license. The number of licenses issued in any year is governed by estimates of the deer population in the state. In years when the deer population is high, more licenses will be issued.

The predator-prey balance is an important influence on wildlife populations. In the absence of wolves, for example, the number of deer can rapidly increase. The high number can exceed the carrying capacity of its habitat, leading to a die-off and the sudden decline in the population.

There are various approaches to managing wildlife. One approach involves the restoration of habitats that have been damaged. One example is the restoration of a wetland. Controlled hunting such as annual deer hunts is another example.

A third example is the monitoring of species and the use of the estimated numbers to decide if the species is under threat. Another approach is to review development planned for a region to determine if it could damage vulnerable land or disrupt migratory routes, as two examples, and, if so, to modify or cancel the development.

2.8. THE IMPACT OF POPULATION GROWTH ON WILDLIFE

Few years back, there was a label which became widely popular, it said: "At least the war on the environment is going well." This irony was used to expose or criticize people of that time for their stupidity.

But the proposal that we have pronounced war on the environment, that we have announced war on millions of other species, is distressingly close to the truth. While the human race has been growing at a rapid rate, and they have increased their presence into every corner of the globe, pretty much everything else is in retreat and decline.

Some of the examples of just how successful the war on everything else has been are discussed as follows:

Several thousand years back, the mass, the weight, all the population on the earth, plus all our pets, plus all the livestock that we maintain to feed ourselves, was 0.1% of 1% – one-tenth of one percent-of the mass, the weight, of all the mammals on the earth. The rest of the mammals-for instance, tigers, elephants, and rhinos and kangaroos and whales etc.,- made up 99.9% of the mass of all the mammals on the earth.

By few decades back, humans, our livestock and our pets had surged from 0.1% to 10–12% of the mass of the mammals of the earth.

Now, the population on earth, our livestock and our pets make up 96% to 98% of the mass of the mammals of the earth. The poor tigers, ancient elephants, whales, rhinos, and kangaroos and all other species of the mammals have reduced from 99.9% to just 2 to 4%.

Since the Year 1500, across the entire world, it was observed that more than 150 species of birds have become extinct. 12% of the species is now at the verge of extinction across the globe, and 220 species of bird are critically endangered.

Out of the widely known European birds, almost half of them are in the decline phase, and 20 widely known North American species have left to 50% in number in the past 40 years.

One in every two mammals are dwindling in number, and nearly one-fourth of the species is at risk of extinction. In addition, it was also estimated that more than 3000 species are critically endangered, which embrace land icons such as the Sumatran orangutan and the African mountain gorilla, as well as sea icons such as dolphins, whales, and seals.

In Australia 59 species-which is almost 20% of their total species, which are threatened; while the mountain pygmy possum is lessened to 2000 and the Tasmanian Devil has undergone a 60% plunge in numbers in 10 years.

It was also noted that Australia's coastal shorebirds are experiencing freefall situation. Between the year 1983 to 2006 the Sooty Oyster-catcher, which is a resident shorebird, plunged by 81%, while the Whimbrel, which is a migratory shorebird, plunge by 73%.

Between the year 1980 and 2000, almost one-third of Australia's bird species had been sank, examples include:

Banded Lapwing (60% fall), Emu (50% fall), Wedge-tailed Eagle (40% drop), Diamond Firetail (40% drop), Black-Shouldered Kite (40% fall), Superb Lyrebird (30% fall), Gang Cockatoo (37%fall), Brolga (27% fall).

Almost 45 species of Australian birds are on the verge of extinction because of climate change. Some of its examples comprise the Wedge-tailed Shearwater waters of the Great Barrier Reef that are too hot for its preferred food; the Fairy Tern, whose nests were being abolished by foxes because of consistent decline in levels of water in the drought stricken Murray, and the Mallee Emu Wren has compromised with almost 50% of its population within the past decade because of the drought and bushfires.

The more the population of humans in this planet, the less the population of everything else.

Some species have been threatened because of the extensive human activity, but the vast majority have not, and many species are now on the verge of extinction.

In December 2005, the USA based National Academy of Sciences came with a surprising report, that because of human activities, the extinction rate of animals was multiplied to 100 times in comparison with natural rates. According to a report by the World Conservation Union, nearly 800 species have become extinct since the year 1500, when accurate records began. The Alliance for Zero Extinction has acknowledged an additional 794 species that are on the verge of extinction.

These species are restricted to 595 sites all over the world; and only 33% of them have legal obligations, and most are existing at those areas where humans are densely populated, almost three times the global average. The country that has the poorest record across the world for species extinction is Australia. In Australia, it was noted that 23 bird species, 27 mammal species, and 4 frog species have become extinct over the past 20 decades.

In addition, the numbers for other Australian species are also not good. According to the Humane Society International, because of the recent land clearing activity in Australia, there was an extinction of almost 4 billion birds, mammals, and reptiles since the year 1972.

In November 2003, the Swiss-based World Conservation Union represented the list of 12,259 different sorts of animal, plant, and water life as critically endangered. It also observed that famous islands such as Hawaii, Galapagos, and the Seychelles are becoming aesthetically and ecologically barren as an outcome of extensive human activities.

Brazil, Indonesia, Peru, and China have the maximum number of endangered mammals and birds, while plants are mostly on the verge of extinction in countries like Malaysia, Ecuador, Sri Lanka, Brazil, and Indonesia.

Some of the activities that are responsible for this repercussion are forest clearance, Industrialization, and tourism.

In Venezuela and Columbia, the population of spider monkeys has been gradually shrinking into smaller and smaller areas by, agriculture activities, urban growth and cattle ranching. The population of the Giant Catfish in the region of Southeast Asia's Mekong River has been fallen by 80% since 1990 because of over-fishing and blocking of dams which is its migratory routes.

Reg Morrison put it like this: "the 300 million tons of humanity that the Earth currently supports has a craving so insatiable that the planet and its biota can fulfill people's demands only by divesting itself of vast numbers of other energy consumers."

The biologist Edward O. Wilson estimated the contribution of humans for extinction stood at between 10% and 20% of Earth's prehistoric inventory of species.

While the standard 'background' extinction rate is about one species per million species each year. Because of extensive human activity, the extinction rate has been multiplied 100 or in some cases even 1,000 times over this level in the rainforest by the reduction in area alone.

According to a report published by United Nations Food and Agriculture Organization, the earth has shrunk its last 5% of tropical forest cover and is evading that at a frequency of over 200,000 square kilometers each year, with the likelihood of doubling the rate.

As the world has entered the twenty-first century with slightly more than 10% of its original forest cover intact. According to Roger Lewis and Richard Leakey, widely recognized anthropologists, it was estimated that the entire forest cover will be more probably devastated by the year 2050.

In Australia, various things are threatening our birds, animals, as well as plants, but none more so than the loss of vegetation cover and the loss of habitat. If the population of this planet really wants to save plants, birds, and animals, there is a need to put an end to habitat loss. Who are responsible for this significant habitat loss?

Well that is obviously every citizen of this world. It is noted that the significant increase in population is the key driver behind the habitat destruction. The hard truth is that is the environment voluntarily groups have been very hesitant to raise the issue of population over the years. One cannot deny the fact that the issue of population is fraught with racial and religious overtones.

2.9. WILDLIFE POPULATION TRENDS IN PROTECTED AREAS

Figure 2.7. Determining the population in a particular area.

Source: Image by National Park Service.

Everyone needs to admit that our biodiversity is in crisis. One possible action taken by the concerned authorities in response to global biodiversity decline and the accompanied threatening processes, has been the setup of protected areas (PAs). PAs triggered the initiative of adopting most global and national conservation strategies, which is aimed at protecting and securing at least 15.4% of the global land surface area.

It is expected that the importance of PAs will likely to increase in future given the present Convention on Biological Diversity goal to expand global land coverage of PAs to 17% by the year 2020. Ensuring that biodiversity is preserved within-PA boundaries is therefore central to accomplishing the objective of global conservation.

Despite a common goal of PAs to conserve wildlife populations within their boundaries, many PAs are witnessing unwanted wildlife population falloffs. Across the world, the population of wildlife changes in PAs are patchily documented, non-measurable, and scantily unstated.

While a widely believed notion is that PAs are effectual at sustaining wildlife populations within their boundaries, this hypothesis has not been broadly verified. Equally, the opinion that some PAs are consistently declining, or at least not performing properly, has triggered calls to drastically change both PA management and conservation related decision-making strategies and, and accentuated the necessity to ensure that PAs are adequately controlled over a longer period of time.

It is of utmost importance to understand how well PAs are conserving wildlife, and to take an assessment of current prevailing conditions and barriers to effective conservation. By gaining an insight into characteristics of those PAs more likely to preserve wildlife populations, it is more likely that new PAs are established in a more spatially and economically competent configuration and amplify biodiversity outcomes within resource constraints.

Without a better understanding of those factors that contribute to wildlife results in PAs, then their design, assortment, and organization are likely to remain sub-optimal.

2.10. THREE MAIN FACTORS THAT CONTROL POPULATIONS

1. **Natality:** One of the basic and important principles involved in assessing the dynamics population of any species is evaluating the number of animals innate every year. The total number of young one of a particular species born in a certain period of time is referred to as the natality or birth rate. It is generally seen that most of the big species reproduce once in a year, therefore biologists are mainly interested in looking for the young ones which are born to each female in a particular year.

The average number of young ones born in a specific year is normally used to make a comparison between populations birth rates of the same species. It is difficult to measure birth rates because they vary from year to year dependent on number of factors like number of reproductive females in the population, prevailing environment or climatic conditions, natural food conditions during the previous year, the density of population, and mating routines of the species. It is generally seen that wildlife biologists collect the data related to birth rates from both harvested animals and animals on which radio transmitters were embedded during the research projects. This will help in assessing the number of births that bears and deer would give by inspecting ovaries from harvested animals and carefully examining placental scars or developing embryos.

This technique was earlier widely popular and was mainly used for deer in West Virginia, but now is not much in use because the greatest number of deer handled by biologists at check stations have been field dressed and their entire organs had been removed. In addition, it was also seen that wildlife biologists assess the reproductive tracts of harvested black bears in order to take count of either placental scars or embryos.

If there are presence of embryos in the reproductive tract, then it is more likely that bear was bred in the year of harvest and would give birth to young ones in the month of January or February of the next year. While, on the other hand, if there are the presence of placental scars in the reproductive tract, it is highly likely that the bear would have given birth in the year of harvest. Female bears usually have cubs every alternate year.

Although in most countries, it is not legal to harvest a female black bear followed by cubs, females occasionally do not possess litters of cubs to term. Suppose, if a female due to some reasons lost her litter after the breeding season, then she would not be pregnant but the presence of placental scars would be clearly seen.

2. Mortality:

Figure 2.8. Checking the mortality rate among wildlife population.

Source: Image by Health.mil.

One of the other key factors that affect the population dynamics is mortality, which refers to the number of animals that die at a particular period of time (deaths per year). It is often observed that people ask questions from wildlife biologists that how long can a particular species live.

The dominant problem about revealing the data of longevity is that the ages regularly come from animals in captivity which live farm more when compared with animals living in natural environments.

Wildlife biologists are mainly interested in knowing the possible reasons for death and the percentages at which these mortality events occur. Some high-quality data associated with mortality rate can be gathered by finding the age of animals when they killed by hunters.

It is easier to collect the information about age at death from hunter-harvested animals as a large number of data can be easily available at a cheaper rate during a limited period of time. The collected data then can be used into what is called a life table (or in some countries a death table) to "back-calculate" population size in preceding years.

As a population ages, there are fewer and a smaller number of individuals in the oldest age classes due to death. Life tables are very useful for biologists as it allows them to compute survival rates of populations and study the interrelation between the birth and death rate.

Actuaries make use of human life tables in order to compute life insurance premiums! Wildlife biologists, on the other hand, make use of animals' data such as their birth rate and mortality rate during research projects with the help of animals wearing radio transmitters to add or verify to the data generated by life tables.

Animals on which radio transmitters have been embedded make it easy for researchers to compute the real number of young ones, which can vary from the number of young ones estimated by observing ovaries. Moreover, animals that wear radio transmitters make it viable for biologists to assess natural mortality, that would not be computed with a life table.

3. **Emigration and Immigration:** Emigration and immigration are both interested terms and mostly used together. Animals that leave their living area, depart from their population and do not return back are called emigrants. Immigration refers to one side movement of the entity into a new area or population. It is widely assumed by wildlife biologists that the effects of immigration and emigration neutralize each other out, but this is not always the case.

The concept of dispersal is associated with both immigration and emigration. Dispersal can be defined as the movement of an animal from one place to another, from the place where it was born to the place where it was shifted.

Dispersal generally refers to a steady occurrence in populations of wildlife and usually arises more frequently when there is high density of population. In most of the wildlife species, males mostly have tendency to disseminated more comparative with females.

High dissemination tendency is conducive to higher mortality rates for males. According to researchers, it was predicted that males disperse farther than females as an adaptation that prevents inbreeding.

2.11. CONCLUSION

In the end, it is concluded that the fluctuation of the population in wildlife is not a new situation. It has been in existence for the past many decades. However, there is a need to shift some focus on this side as if proper action would not be taken, then there has been significant decline in the population of endangered species.

In addition, there is need to find a balance between plants and animals as both are dependent on each other for their existence. Variation in one's population hay have subsequent impact on the other one. Therefore, it is the duty of ecologists and wildlife experts to engage in planning and decision making to protect the interest of plants and animals and ensure their conservation and sustainability over a prolonged period of time.

REFERENCES

1. Encyclopedia Britannica, (2013). *Population Ecology-Life Tables and the Rate of Population Growth*. [online] Available at: https://www.britannica.com/science/population-ecology/Life-tables-and-the-rate-of-population-growth (accessed on 21 August 2020).
2. Encyclopedia Britannica, (2020). *Population | Biology and Anthropology*. [online] Available at: https://www.britannica.com/science/population-biology-and-anthropology (accessed on 21 August 2020).
3. Encyclopedia.com. (2020). *Wildlife Population Management | Encyclopedia.Com*. [online] Available at: https://www.encyclopedia.com/environment/energy-government-and-defense-magazines/wildlife-population-management (accessed on 21 August 2020).
4. Nhpbs.org. (2020). *Population | Wildlife Journal Junior*. [online] Available at: https://nhpbs.org/wild/population.asp (accessed on 21 August 2020).
5. Population Media Center, (2011). *Impact of Population Growth on Wildlife-Population Media Center*. [online] Available at: https://www.populationmedia.org/2011/07/15/the-impact-of-population-growth-on-wildlife/ (accessed on 21 August 2020).
6. *The Ups and Downs, Hows and Whys of Wildlife Populations*. (2020). [ebook] Available at: https://www.wvdnr.gov/wildlife/magazine/Current/Wildlife%20Populations.pdf (accessed on 21 August 2020).
7. www.nature.com. (2016). *Wildlife Population Trends in Protected Areas Predicted by National Socio-Economic Metrics and Body Size*. (2016). [online] Available at: https://www.nature.com/articles/ncomms12747 (accessed on 21 August 2020).
8. Yarrow, G., (2009). *The Basics of Population Dynamics*. [ebook] Available at: http://uaex.edu/environment-nature/wildlife/youtheducation/TR%20Basics%20Population%20Dynamics.pdf (accessed on 21 August 2020).

3

Factors Responsible for Declining Wildlife Populations

CONTENTS

The earth's ecosystems have undergone a sea change and have been impacted negatively over the years, due to the actions of humans. As a result, the wildlife populations have declined drastically and are faced with the threat of extinction.

The flocks of birds that thrived in marshy swamps and lagoons have their numbers reduced dramatically, the sounds of the croaking of frogs and chirping of crickets in late nights are now a distant memory and large herds of bison roaming freely in open prairies has now been limited to pictures and images. The numbers of large animals like elephants, giraffes, and whales have also reduced, which in some cases are facing the threat to their survival.

In recent years, we have seen many pictures and also read news items regarding whales, dolphins, and other marine species that are getting washed-up dead-on ocean beaches, sometimes in considerable numbers Also, many reports have come up regarding drastic depletion in fish populations.

In this chapter, the various factors responsible for dwindling wildlife populations including how these factors contributed to their decline shall be discussed.

3.1. INTRODUCTION

Over a period of time, growth of the human population and the increase in human activity has severely and badly affected the forest cover as well as the wildlife. It is a fact that with the beginning of the industrial revolution in the world, especially Europe in the eighteenth century, a good number of wildlife population (including animals and plants) had been extincted.

To meet their needs for development, humans have not only exploited the natural resources of the earth, they have also encroached the landscapes and waterscapes, where wildlife thrived. This has resulted not only in deterioration of the natural environment but has also severely threatened the earth's ecosystems, by means of unhealthy production and consumption.

Consequentially, during the last three hundred years, not only a good number of wildlife species have gone extinct, the individual count of numbers of animals and plants across the various species has also declined; in some cases, drastically.

This has resulted in adverse effects on many aspects of wildlife, most importantly like genetic variation and biodiversity. It is a matter of concern that if ecosystems are severely depleted, their ability to sustain and replenish the various species and meet the needs of humans is threatened severely.

3.2. DEVELOPMENT AT THE COST OF ECOSYSTEMS

The benefits of economic growth and development, achieved by the developed nations has come with a price. For achieving their economic and industrial growth, the developed countries have not only exploited the natural resources and ecosystems within their own national boundaries, but also by exploiting the natural resources and ecosystems of the underdeveloped countries.

Presently, the biggest decline in animal and plant life and damage to the ecosystems is occurring in the developing countries which have low incomes and high populations. The decline in the wild is developing countries is well comparable to the similar losses faced by developed nations long before.

For example, between 1990 and 2008 European Union countries imported about one thirds of the products that are responsible for deforestation, which is timber, beef, and soya.

3.2.1. Are We Facing New Process of Mass Extinction?

Experts in the field of wildlife often debate whether we are now facing to face with the new process of mass extinction or apocalypse akin to the ones that the world had seen over the millennia. Even if this is not going to happen, it is a known fact that today that thousands of wildlife species come under the endangered category. Most of plant and animal life on both land and sea have been severely reduced in numbers and this is a matter of grave concern.

As per a report of the International Union for Conservation of Nature (IUCN), since the beginning of the sixteenth century around 850 species have disappeared. It is felt that many more specified had disappeared from the face of the earth before they could be identified by the scientists. Many more species face the threat of extinction in coming period. It is estimated that around 33% of amphibians and 20% of mammals are severely endangered and may be extinct in the coming few decades.

Some people have argued that the present decline in the population of animal species is a part of the natural process as was the disappearance of the species in the past. But the fact is, that the mass extinction presently faced by the species is a result of the human intervention, whereas, in the past, the process of global mass extinction of species on earth was due natural calamitous events.

According to some experts, the phenomena being experienced today have a lot of similarities with the catastrophic event sixty million years ago, that wiped out the dinosaurs from the face of the earth. Mass extinction of dinosaurs occurred when earth resources already stressed by the presence of large size reptiles and climatic changes were struck by asteroids.

Figure 3.1. *Marmoset white tufts are gradually becoming extinct as primates.*

Source: Image by Needpix.

The process of mass extinction, which is being faced by wildlife species is because of the intervention by humans. The negative effects caused by the actions of humans are multifold on other living species. As a result of humans' activities, the total number of wildlife species, as well as the population amongst several species, has undergone considerable reduction. Many species have disappeared from the face of the earth and many more are facing the threat of extinction.

The crises of mass extinction are not only being faced by wildlife species existing in terrestrial biomes like tropical forests, but also by species living underwater in marine biomes. Even today we do not have the complete knowledge regarding the number of species that are present in the ocean, as many such species have not yet been discovered.

Therefore, it is not known, how many oceanic species have already disappeared and how many are faced with the threat of extinction. Further, not only the species are being lost, but certain specific unique type of sub-species within a species has disappeared, before they could be studied (e.g., a type of salmon). As a consequence, the extinction of such species and loss of their genetic variation leads to loss of biodiversity, which is an irreversible process.

Current estimations of the fish populations cover only 20% of the fish stocks, so the correct state of the total fish population is still unknown. Some of the recent studies have indicated that approximately 75% of world's commercially fish stocks are over harvested and face the risk of extinction.

Of late, there have been reports of new species of animals and plants getting discovered in tropical regions across the globe as well as some areas of oceans, that give us a sense of hope and possibility. But it is cruel fact that enormous areas of natural forests that were home to varieties of flora and fauna species are being destroyed unabatedly every year.

3.3. DRIVERS OF EXTINCTION PROCESS

The drivers for the process of extinction are briefly given as mentioned below:

1. **Habitat Loss:** Cutting down trees, Filling of wetlands, dredging of rivers and mowing of fields are some of the actions that lead to habitat loss.
2. **Habitat Fragmentation**: Terrestrial wildlife is cut up into fragments as a result of some development activity or construction of roads across the wildlife habitats. Similarly, aquatic wildlife is fragmented through the construction of dams or some water diversion. Such fragments may either not be large enough or connected to support the species for breeding and food.

Loss and fragmentation in habitat also cause difficulties to migratory birds in finding places to feed and rest along the route, resulting in change in their migratory patterns.

3. **Habitat degradation**: High pollution, invasive species and disruption of ecosystem processes are some of the factors that cause habitat to be so degraded that they are no longer able to support the native wildlife.
4. **Climate change**: The weather and temperature patterns altered due to climate change have a serious impact on flora and fauna.

Due to temperature rise the numbers and variety of species that define biodiversity reduce greatly.

Burning of fossil fuel for energy and animal agriculture along with deforestation are the major factors contributing to global warming.

Figure 3.2. Loss of habitat is a major reason for wildlife decline.

Source: Image by Wikimedia Commons.

Livestock accounts for up to around 18% of human induced greenhouse gas emissions from cattle waste, the fertilizer production for feed crops; general farm associated emissions and the process of growing feed crops.

A study shows that the consumption of dairy and meat products by people increases with their income. In developed countries, meat consumption is almost double of meat consumption in developing countries. A data indicates that the worldwide meat production has increased by 40% in the last 40 years.

3.4. THE ROLE OF HUMAN ACTIVITIES IN DECLINING WILDLIFE POPULATIONS

Human activities are the single major cause for depletion in global wildlife population. Species are disappearing from face of the earth due to loss and degradation of their habitat, overexploitation of wildlife, pollution, and climatic changes.

According to a recent report of World Wildlife Fund (WWF), over the last 40 years, species counts have been reduced by an average of 60%. WWF experts tracked populations of more than 4000 wildlife species (mammal, bird, fish, reptile, and amphibian) and found an alarming decline in these species between 1970 and 2014.

This decline in populations of species in considered driven by human activities as the rate of extinction was multifold, as compared to times when human pressures were considered as a factor. The rapid increase in human population growth rate from twentieth-century onwards led to quick depletion in earth's ecosystems, with forests, wetlands, mangroves, and coral reefs all getting reduced. Before the twentieth century, the consumption of resources was lesser that the Earth's rate of renewal.

In the last fifty years, it is estimated that 20% of Amazon rain forests, considered as lungs of planet Earth have disappeared and half of the shallow water corals that existed on earth have been lost in last three decades.

These changes driven by human beings are highly invasive and may be leading towards mass extinction of species from face of the earth. As per WWF report, "Earth is losing biodiversity at a rate seen only during mass extinctions. This situation may further get aggravated as per the report as "people are responsible for releasing 100 billion tons of carbon into the Earth's system every ten years."

WWF called for urgent action on climate change, endorsing a UN report that calls for the requirement of rapid, far-reaching and unprecedented changes in all aspects of society, in wake of the potentially irreversible threat posed by climate changes to earth.

Further, as per WWF, some sort of global actions, similar to those laid out in Paris Climate Agreement are required urgently, which can set ambitious targets to reverse the process of mass extinction of species. Noting that all hope is not lost, the WWF report says "We are the first generation that has a clear picture of the value of nature and the grave situation we are facing" and "we are the last generation that can do something about it."

3.4.1. Catastrophic Animal Decline and the Living Planet Index

In one of the *Living Planet Report* of WWF for Nature, a report regarding sharp decline in animal population all around the world was published, which was vastly misinterpreted with news spreading all around that the world has lost 60% of its wildlife population over a period of last 40 years. In reality, the situation is not that bad although it is deeply concerning.

The Living Planet Report had analyzed the trends of population decline in animals globally in terms of Living Planet Index (LPI), for the diversity and abundance of animals globally. If the global score is steady or increasing, the animal populations are generally good and if it is decreasing it shows that the population is on decreasing trend.

Thus, the Living Planet Report, in fact, stated that the LPI had sharply decreased by 60% in the past 40 years between 1970 and 2010.

The LPI is a measure of the state of the world's biodiversity based on the population trends of vertebrate species from terrestrial, freshwater, and marine habitats. The LPI has been adopted by the Convention of biological diversity (CBD) as an indicator of progress towards its 2011–2020 targets 'to take effective and urgent action to halt the loss of biodiversity.'

The LPI is worked out by considering the data over a wide spectrum of species that have entirely different conservation status as well as lifestyles. Different weightages are given to different species. As an example, the weightage given to smaller size species that have faster and frequent reproduction rate and have a very high population like Eurasian pygmy shrew (a type of mole) is far lesser than large animals or vertebrates which have infrequent reproduction and less population, like tiger or an endangered species of rhinoceros-like Sumatran rhino.

In simpler terms that means the loss of one rhino or one tiger is considered a much greater loss in comparison to the loss of a few thousands' pygmy shrews. Therefore, coming back to a figure of 60% loss of animals as given in Living Planet Report, it implies that some animal populations lost more than 60% of its individuals and some lost far less.

In other words, as per the definition of LPI, the report stated that the population of vertebrates on an average declined by 60%. That does not mean that the total population of animals across all species declined by 60%. This point is also clearly mentioned in the report but nevertheless, the average as per LPI indicates the catastrophic global trend in decline of animals.

To illustrate the above, let us consider the following two examples, one with multiple species and another within a single species:

Let us consider an imaginary population of 50 lions, 200 falcons, and 10,000 squirrels. If the population of lions declines by 90% to 5 lions, the population of falcons declined by 80% to 40 falcons, and that of squirrels declines by 10% to 9000. In this case, although the average decline of imaginary population by 60%, but quantitatively the total decline is only 12%.

Taking another imaginary case of single species of population of gray wolves, let us say that the population of gray wolf has declined by an average of 60%. Considering that gray wolf is found in different pack sizes and at different locations, this does not imply that 60% of total gray wolf has perished. It means that at a certain location gray wolf packs have suffered tremendous losses or have gone into extinction locally, whereas, at other locations, the decline in their population may not be so sharp.

3.5. ANIMALS PROTECTED BY THE ENDANGERED SPECIES ACT

3.5.1. Ecosystem Services

For decades, there is a fight going on between the economists and ecologists, and the reason behind that fight is to find how to quantify the value of ecosystem services, which means the goods and services which is provided by nature, from bee-assisted crop pollination to water filtration courtesy of mollusks. The value of ecosystem services is fixed at $125 trillion by The Living Planet Report, which is just a few trillion shy of the world's total GDP.

However, it is said that the numbers may be controversial, as how to assign such values is not agreed by the people. Even nature does a lot of the heavy lifting for everyone which has to be admitted by Wall Street. For a third of all food production animal pollinators are responsible, and it is said that to recycle soil everyone is mainly depends on cud-chewing animals. For the next cancer treatment, medical researchers look to elusive rainforest critters but at the same time birds, bats, and reptiles rein in disease-spreading mosquitoes.

The fate of a single animal population can have far-reaching and unexpected consequences due to intricate food webs.

3.5.2. Census Findings

The frustration of working on a biodiversity census by comparing it to other nature-monitoring projects with extensive infrastructure, like weather prediction is explained by the University of Maine biologist Brian McGill.

Forecasting weather is not an easy task to do, so in the U.S there is an agency name National Weather Service which spends billions of dollars a year to make accurate weather forecasts.

Spending billions of dollars here means investing in ground weather stations, ocean buoys, radiosonde balloons, and satellites, and this done only to get the best possible measurements of the current state of the weather. It is said that for biodiversity there is nothing equivalent.

The WWF keeps tabs on 4,005 species of vertebrates in 16,705 distinct populations through its Living Planet Reports. As there are 63,000 total described species of vertebrates so this figure is dwarfed by it. And the scientists have a mindset that they have only documented a fraction of what's out there.

Populations serve as a point of reference for all the species on which there is no data available, as they are strategically scattered across continents and biomes. For example, to get a precise headcount on smaller animals like ants is very hard.

And on the other hand, scientists get a good sense of how the local ecosystem is holding up because at the same time it is keeping up with birds and mammals that feed on them in the same neighborhood.

According to the report, freshwater, and tropical ecosystems are the neighborhoods which are in the biggest trouble. As they both are considered as the historical hotspots for biodiversity, which is supporting an unusually high concentration of species. These once-rich habitats have whittled away by the overexploitation, climate change, and pollution.

By 83% the average LPI score for freshwater populations has plummeted. This means a population of water-dwelling animals has declined by this amount on average. And the LPI has fallen by 89% in the neotropics—the Caribbean, Central, and South America, and slivers of Florida and Texas.

3.5.3. Ecosystem Politics

On a global scale stemming the tide of extinction will require cooperation, so by an international backsliding on environmental regulations scientists have been alarmed.

There is an act in the U.S Endangered Species Act in which the Trump administration continues to push to roll back elements, and it is said to be the United States' bedrock environmental policy. And earlier this week, the trade of rhino horns and tiger bones is legalized by the Chinese government, this said to be a move conservationist have decried.

At the same time to develop the Amazon to bolster his country's economy Brazil's president-elect Jair Bolsonaro plans, as everyone knows that biological diversity loss and ecosystem collapse as one of the top 10 economic threats in 2018, which is highlighted by the World Economic Forum's Global Risk Report.

3.6. FACTORS AFFECTING WILDLIFE LOSS

Wildlife refers to undomesticated animal species that survive in natural, wild areas throughout the world. In order to sustain healthy ecosystems, a delicate balance must exist among all living things.

Unluckily, by human involvement, that balance has been thrown off Mammal, bird, reptile, fish, and there is a decline in the population of amphibian by 52% during the last 4 decades according to the WWF. The forces behind this unprecedented destruction are many.

3.6.1. Human Land Use Competing for Space with Wildlife

It is said that land use/cover change is considered as a major driving force of habitat modification. It has important implications for the distribution of wildlife species and ecological systems.

There is a successful coexistence between wildlife conservation and local uses, and conditions for the environmental, social, and economic sustainability of such coexistence is a question that has not been adequately addressed.

It is said that rapid land use/cover change is said to be dynamic, widespread, and an accelerating process, which is mainly driven by natural phenomena and anthropogenic activities, which in turn drives changes that impact humans.

Figure 3.3. Humans have been exploiting the land for their benefits and uncontrolled wants.

Source: Image by Flickr.

It is said that the rapid land use/cover changes are unprecedented in human history and especially in developing countries it is continuously changing ecosystems, thus threatening sustainability and livelihood systems.

With the increase in the human population, for space and resources biodiversity is facing widespread competition with humanity and as a result, between the need for biodiversity conservation and economic development conflict is increasing.

Protected areas (PAs), to protecting wildlife species the cornerstone of modern biodiversity conservation go some way. However, they don't resolve the conservation-development conflicts since they do not always exclude destructive human impacts nor they completely protect the biodiversity.

Only a part of an ecosystem or species range is termed as the PAs which tend to be protected, and it is said that there will be an increase in conflict with the local people with the wildlife dispersal from such areas.

To achieve sustainable benefits from wildlife, even when alternative forms of land use are there, such as wildlife tourism. It is said that they are implemented in such a way that sustainable benefits can be achieved and conflict may still remain.

The main challenge is to allow economic development so that mutually sustainable benefits can be derived and at the same time another challenge is to identify strategies that ensure conservation.

On the basis of informed decision-making and consensus among various stakeholders, it is considered as the most difficult issue especially in developing counties and it requires detailed understanding of issues, careful monitoring, and adaptive management.

3.6.2. Impacts of Air Pollution and Acid Rain on Wildlife

Animals are exposed to air pollutants via three pathways:

1. Inhalation of gases or small particles;
2. Ingestion of particles suspended in food or water; and
3. Absorption of gases through the skin.

In general, there are some soft-bodied invertebrates (e.g., earthworms), or animals with thin, moist skin (e.g., amphibians) are affected by the absorption of pollutants. It is said that an individual's response to a pollutant varies greatly and depends on the type of pollutant involved, the duration and time of exposure, and the amount taken up by the animal.

There are some other things that are considered in playing a role in its response are the individual's age, sex, health, and reproductive condition. It is said that between animal classes, species, and even genotypes there is a great deal of variability, in terms of the level of tolerance to a particular pollutant.

There are three broad categories in this section in which the pollutants of concern have been divided, gases, such as ozone and hydrogen sulfide; non-acidic particulates and toxins, like metals, fluoride compounds, and organic and synthetic chemicals; and acidifying agents, specifically nitrates and sulfates.

1. **Gaseous Pollutants:** From the industrial process's volatile organic compounds and nitrogen oxides are emitted. As, in the atmosphere, in the presence of sunlight to form ozone it undergoes chemical transformation.

There is something which primarily affects the respiratory system is ozone, sulfur dioxide, and nitrogen dioxide. It is said that birds are considered as the more susceptible one to gaseous pollutant injury than mammals due to their higher respiratory rates.

Figure 3.4. Gaseous pollutants prove extremely dangerous for the wildlife.

Source: Image by Pexels.

2. **Non-Acidic Particulates and Toxins:** Particulates are the one in which the number of air pollutants are categorized. Smelters help to emit heavy metals (e.g., lead, arsenic, and cadmium); from aluminum reduction plants and coal-fired power plants fluoride is emitted in both gaseous and particulate form; and by resource recovery facilities dioxins, furans, and mercury are emitted.

Circulatory, respiratory, gastrointestinal, and central nervous systems of animals may get affected by the metals. Kidney, liver, and brain are the organs which are targeted. Changes in birth, growth, and death rates are the cause can be affected as metal contamination to the entire population.

Gross malformations of bones and teeth are the thing which is caused by the fluoride poisoning, or fluorosis. Gaseous fluoride is taken up by the plants take up and plants store it in their tissues, and are deposited on leaf surfaces in particulate form of fluoride and stay there until washed off.

For exhibiting symptoms of fluoride poisoning herbivores are the one which is best known for it. However, fluoride, which is, in turn, passed on to the animals that eat them earthworms and other soil invertebrates also accumulate.

Dioxins and Organ chlorines are the organic and synthetic chemicals which affect wildlife. By concentrating in body fat dioxins bioaccumulate or build up in the body and it is said that they are resistant to biological breakdown.

Up to the five times, the concentration found in the soil is accumulated by the earthworms, as it is shown in one of the studies, even though it is known that this dose is not deadly to the worms.

Since earthworms are a major source of food for a number of bird and small mammal species, so this non-lethal could have a strong ecological implication and many of which have exhibited carcinogenic, reproductive, and immunotoxic effects after exposure to low levels of dioxins.

3. **Nitrates and Sulfates:** As a result of fossil fuel combustion undergo chemical transformation in the atmosphere sulfur dioxide and nitrogen oxides emitted, and when they get dissolved in precipitation known as acid rain than they occur as sulfate, nitrate, and hydrogen ions. Sulfate and neutralize acidity can get adsorbed by the well-buffered soils.

This adsorption will result in stream water and soil water composition being maintained in a range acceptable to organisms. Even it is said that well-buffered soils adsorption capacity is limited, and the supply of base cations in the soils that buffer these inputs is depleted because of the long-term deposition of acidic compounds.

Delay in the acidification of surface waters once saturation is reached in sensitive watersheds because of the result created by the build-up of sulfates and nitrates in soils.

In a National Acid Precipitation Assessment Program (NAPAP) report all the effects of decreasing pH on aquatic invertebrates and fish have been summarized. With some species affected at pH levels near 6.0. insect taxa differ greatly in their response to acidity.

It is said that acid-sensitive species are replaced by acid-tolerant ones in the early stages of acidification. However, if more species are lost than pH levels will continue to drop.

In many studies, it has been shown that there is a decline or loss of fish population because of the surface water acidification. It will be impossible for a fish to survive when pH is below 4.5. Scandinavia, Scotland.

Wales and North America are some of the countries in which fish loss is occurring and also there are many other countries too. With an increase in metal availability, being particularly true for aluminum and mercury a decrease in pH is often associated.

To increase fish mortality, decrease fish growth, decrease egg production and embryo survival decreased pH and elevated aluminum has been shown, and which will result in physiological impairment of adult fish.

In general, it is said that adult fish are more acid-tolerant than embryos, fry, and juveniles. Onto fish gills, inhibiting diffusion, and resulting in respiratory stress can be precipitate by aluminum.

It is said that if there is a decline in the amphibian populations than the acid deposition is considered as a possible cause. By the acidic water, the larval stages of aquatic amphibian species are most affected.

There are temporary ponds which are used by many frog species, but these tend to be small and shallow, and their only sources of water are rainfall and snowmelt which are easily affected by precipitation chemistry.

There are some large permanent bodies of water which are used by the frogs for breeding, generally lay their eggs in the summer, so acid pulses from snowmelt is not experienced by the frogs. However, to subtle changes in pH levels than those of species that breed in the temporary ponds, the eggs and larvae of these species are even more sensitive.

As it is said that it is true with fish, when concentrations of metals, such as aluminum, in the water increase the toxic effect of decreased pH levels on amphibians is complicated, but as a general rule, embryos of sensitive amphibian species are killed by water with a pH of 4.5 or lower, while embryos of tolerant species can survive down to a pH of 3.7.

4. **Indirect Effects:** Moreover, it is said that affecting individual animals or populations directly, and with the changes in the ecosystem air pollutants will increase and it will also affect wildlife indirectly. From predators and weather vegetation affords cover for protection and along with this it also provides breeding and nesting habitat, and also serves as a food source.

Therefore, there will be an indirect effect to the animal populations, if there is any change in vegetation. Many studies have found that a preference is shown by the invertebrates for, or are better able to establish them in, air pollution-injured vegetation.

Levels that are toxic to soil invertebrates in which fluoride and heavy metals can accumulate in the soil. More metal-tolerant are the one by which species sensitive to metals are replaced. For example, earthworms, and nematodes seem to be more readily affected by elevated metal concentrations and they are the soft-bodied species.

In forest floor litter decomposition invertebrates play an important role. It is said that mineral release is delayed, as the forest floor litter builds up and it will also result in the reduction in the availability of nutrients to plants. When the quantity or quality of their food supply decreases, then ultimately herbivores are affected.

Although it is said that by water acidification birds and mammals are not directly affected, but if there is a change in the quality and quantity of food resources than they are indirectly affected. There are some birds who find difficulty in living around an acid lake such as the osprey, because there are far fewer fish to be found.

However, it is said that the water of an acid lake is clearer than that of a normal lake, so this is the only reason that diver finds hunting easier in an acid lake. In Scotland, their main food supply which is fish is reduced because otters are quite rare around acidic streams and rivers.

For both mammals and birds calcium is considered an essential element. So, during reproduction an adequate dietary supply is crucial. For the proper formation of eggshells and for skeletal growth of hatchlings birds need calcium, and on the other hand for skeletal development of fetus's mammals need calcium.

Mollusks and Crustaceans are the invertebrate species that contain high concentrations of calcium, as they are very sensitive to pH levels and during the acidification of wetlands, they are the first to disappear.

3.6.3. Invasive Species

There are many other names for Invasive species and that significantly modifies or disrupts the ecosystems as it colonizes such as species, alien species, or exotic species, and any nonnative species. Through natural migration, such species may arrive in new areas, but by the activities of other species, they are often introduced.

Those who are involved in global commerce and the pet trade are known as human activities, are considered to be the most common ways invasive plants, animals, microbes, and other organisms are transported to new habitats.

In new habitats most introduced species do not survive in the extended periods, because to adjust to the challenges posed by their new surroundings, they do not possess the evolutionary adaptations. When species possess a built-in competitive advantage over indigenous species in invaded areas, it means that there are some introduced species which may become invasive.

Figure 3.5. Candle bush is an invasive species in Mexico.

Source: Image by Flickr.

Under these situations, breeding populations can be established by the new arrivals and thrive. This is done specially, in the case of the ecosystem lacks natural predators capable of keeping them in check.

The ecosystem's biodiversity is often reduced when the ecological disruption that tends to follow such invasions and this results in causing economic harm to people who depend on the ecosystem's biological resources.

Prey populations decline over time because invasive predators may be so adept at capturing prey, and from affected ecosystems, many prey species are eliminated. And on the other hand, there are other invasive species which may prevent native species from obtaining food, living space, or other resources.

With time it will be possible that invading species can effectively replace native ones and often forcing the localized extinction of many native species. It is also said that spread parasites and pathogens that may further disrupt invaded areas because invasive plants and animals may also serve as disease vectors.

3.6.3.1. A Global Problem

The species have migrated and colonized new areas since the dawn of life on the earth. It has been observed that in some of the cases, migrating species were not able to establish sustainable populations in their new habitats and they died out quickly.

On the other hand, in some other cases, these migrating species were either incorporated into the current structure of the ecosystem or these species were responsible for changing native food chains by decimating native prey or by outcompeting native competitors.

During the Pliocene Epoch (5.3 million to 2.6 million years ago), one of the most significant species invasions in the history of the Earth took place. It took place after the creation of an isthmus that was linking North America and South America.

There were various predator species that were migrating from North America to South America. These species are believed to have contributed to the extinction of many mammalian species of South America.

Since the emergence, modern humans (that is Homo sapiens) have played an increasing role in species invasions. As an outcome of their colonization of all but the most extreme of Earth's ecosystems and their tendency to change natural environments into urban and agricultural landscapes, modern humans (Homo sapiens) are amongst the most effective invasive species.

Although, modern humans (Homo sapiens) also contribute significantly to the introduction of various species to new regions. It has been observed that tens of thousands of years ago, migratory bands of humans were also accompanied by pathogens, domesticated animals and parasites.

With the increase in the civilization, various exotic plants and animals were brought from faraway lands in order to expand the palettes of consumers. These plants and animals were also brought to serve as curiosities in circuses and gardens.

However, the transport as well as the collection of exotic plants and animals' dates to ancient times, but all the written records of the ecological effects of these exotic species extend back just a few centuries. The Norway or brown rat (*Rattus norvegicus*) is one of the best-known historical examples of such kind of species. This rodent is thought to have originated in northeastern China and it spread all over the islands of the Pacific Ocean.

The accidental introduction of rat was during the voyages of exploration between the late 18th centuries and 19th centuries. This population has established themselves on various Pacific islands, which includes Hawaii and New Zealand. Here, they prey on various native birds, amphibians, and small reptiles.

In addition to it, there were some other introductions at this time. These were deliberate cats, dogs, pigs, and some other domesticated animals were taken to new lands. All these animals after coming to the new land caused the extinction of various species. This includes the dodo (*Raphus cucullatus*) from Mauritius by 1681.

It has been observed that the invasive species occur on all the continents, but Australia and Oceania have been especially hard-hit. It has to be noted that the first wave of invasive species came in Australia and the islands of the Pacific. It came with the European explorers in the kind of feral cats and several species of rat.

Wild rabbits of Europe (that is Oryctolagus cuniculus) were initially inhabited in southern Europe and North Africa. These wild species of rats were then deliberately introduced into Australia in the year 1827.

These species were introduced to Australia in order to serve as a common element for settlers in a new land. In addition to it, all the wild species of rabbits multiplied drastically. These species, over the time, destroyed grazing lands by stripping the bark from shrubs and native trees and eating their leaves and seeds.

The red fox (Vulpes vulpes) is a small predator that is found across much of the Northern Hemisphere. Since its introduction in the 1850s, it caused havoc on marsupials and native rodents. The red fox was brought to Australia in order to help to control the European wild rabbits that are mentioned above.

In addition to it, there were many other species such as the voracious cane toad (Bufo marinus). The native range of this species spans from northern South America to southern Texas. This species is a poisonous species with some natural predators.

This species was introduced into Australia in the 1930s from Hawaii. It was introduced in order to decrease the impacts of beetles on sugarcane plantations.

It has been observed that cane toads are to blame for various problems, like the decline in the population in native prey species (that is bees and other small animals), decrease in the population in amphibian species that compete with them, and the poisoning of all those species that eat them.

In addition to it, a large number of invasive plants were introduced to Australia. Darwin Botanic Garden may have introduced a Giant sensitive tree (*Mimosa pigra*) sometime before the 1890s; unpalatable to most wildlife, it creates vast thickets and disturbs the native wetland ecosystems.

Lantana (*Lantana camara*), Surinam cherry (*Eugenia uniflora*), Cherry guava (Psidium cattleianum), Arabian coffee (Coffea arabica), and the ice cream bean (Inga edulis) are all the invasive species. These species were brought to the new land as ornamental plants or food and escaped cultivation.

The brown tree snake (*Boiga irregularis*) is a native of Australia and Indonesia. On Guam, Saipan, and various other Pacific islands, this species has caused the extinction of various reptiles, birds, and amphibians and two of three native bat species of Guam since its accidental introduction to these islands in the period of 1950s.

Though the snake might have been brought to the islands to control the population of native rodents, it is more possible that the earliest invaders were stowaways aboard cargo ships and military aircraft.

3.6.4. Disease as an Agent of Extinction

Although it has been established by the fossil record that the fate of all the species is to ultimately die out, it is clear from the same record that the rate of disappearance of individual species differs in a significant manner.

As it is already noted that the assumptions about how (as opposed to when) an individual species vanished must be developed retrospectively and inductively.

The essential guideline is that the clear causes of extinction that are diachronic (repeatedly have impact on the species across time) are inherently more plausible as compared to that extinction that is said to have taken place just once or apply to only one taxon.

However, this means that descriptions about the extinction of individual species are not strictly testable but they can be assessed in terms of likelihood. This is the approach that is presently taken by the IUCN and various other conservation organizations that are interested in compiling the statistics of the extinction.

It is considered as an accepted belief in conservation biology that any severe, long-term threat to a species may contribute to its extinction in the end. From this point of view, it is also understood that illnesses that are appearing with very high rates of mortality—as in the case of a highly infectious infection that is recently growing in a population—can also cause outright endangerment.

It is conceivable that a species would undergo extinction in a single epizootic event only under the most extreme conditions. More likely would be the frequent outbreaks over a period of time gradually decreasing the fitness level of the species, with the ultimate disappearance of the species potentially caused by stochastic events (like causally unassociated change in the climate).

There is a way in which this condition can be achieved. This way would be through a stable carrier (that is a species other than the target, staying in the same kind of situations in a similar environment, and in which the disease is in apparent or at least it is sublethal).

The transfer of simian acquired immunodeficiency virus from one species of macaque to the other species is a well studies example. Though this example happened under captive conditions, the frequent or repeated outbreaks of distemper in lions as well as the wild dogs of Africa have long been thought to be because of the transfer from domestic dogs.

It has to be noted that for a disease to really have some serious effect, it would be required for the pathogen to occur in highly lethal, aggressive strains that strongly have an influence on the target species before attenuated strains occur and it becomes common.

There is a requirement of high potential in order to cause infections in a susceptible species which is usually associated with the ability to successfully enter the organism which can be through the major portal, such as respiratory tract, where it is basically can be lodged and can be even transmitted easily.

In order to achieve the hyper lethality and to produce the serious mortality, it is important that all the age groups those are within the species would be probably have to be susceptible. That is not a very old or young, with which death is the usual outcome.

It is considered that in a large-bodied mammal there is a fundamental consideration, that is basically any process that is deleteriously affects the young individuals and will have a pronounced effect on the survivorship. This is because of the lengthy intervals present in the birth spacing.

Lethality which is in the range of 50 to 75% is considered to be extremely high and therefore is extremely unusual, even though it is historically seen in the case of Ebola infections in humans and incase of experimental transmission studies from the pigs to macaques.

There may be high percentages that may be achieved in the rinderpest outbreaks that were among the East African bovids in the early 20th century. Even though, the quantitative data on this is largely lacking.

There is an important issue, whether the pathogens are causing this level of lethality that could maintain themselves in nature along with the enough to seriously imperil a species.

There is a possible outcome that can be with the hyper-lethal infections which are producing a rapid, fetal outcome that is affected by the populations and would be reduced to a small number of a widely dispersed or relatively or completely immune individual.

Under such circumstances, it is believed that the epizootic would be necessarily abated, as it basically ran out of a new host which is leading towards the conclusion that the exceptionally lethal diseases cannot be indefinitely maintained within a population or a species that are under normal circumstances.

Therefore, if the reservoirs are existing from which the pathogens can be repeatedly emerging, then in the principle epizootics might resurge year after year until the population size was reduced to below viable level that is approximate 50 to 500 individuals.

At this point, there might be stochastic effects that might intervene and can lead to the complete loss of the species. among birds, there are severe impacts of avian malaria on Hawaiian honeycreepers which is also pertinent.

Even though, there are number of honeycreepers species that survive at high elevations, which is above the limit at which induced Culex mosquitos can often survive. There is also presence of multiple adventitious threats like deforestation and competition from the invasive species which often adds their endangerment pictures.

3.6.5. Extinction and Infectious Disease in Mammals

It is believed that there are several current examples of mammalian species those are already under threat for the various reasons of being severely impacted by the infectious diseases. These diseases include canine distemper in the black-footed ferrets and lions, transmissible facial tumor disease in Tasmanian devils, Ebola, and Marburg hemorrhagic diseases in anthropoids, where none of them has resulted in the extinction.

The possible example that exists in the mammalian fossil record, which can be although compelling in some of the instances, which also lack adequate corroboration for the multiple reasons. It is considered that there is only one study that can be considered to meet the appropriate retrospective criteria that is used for identifying the disease as the primary cause of the extinction at the species level in any of mammals.

This study is involved in an investigation of the disappearance of two endemic murid species on a Christmas Island (Indian Ocean) at the beginning of the 20th century. The Christmas Island extinction can be usefully studied at all which is due in a large measure to the work of the biologist Andrews.

Later, the parasitologist Durham, who basically spent considerable lots of time on the Island and hence recorded may valuable observations on the rats before and during their disappearance. It was considered Christmas Island was evidently uninhabited by humans till the last quarter of the 19th century.

With the discovery of the phosphate the deposits seen on the island, its exploitation often becomes inevitable. The Christmas Island was first formally annexed by the United Kingdom in the year 1888. There was an immediate result that often increased the ship traffic. As a result, in the year 1899, the black rats known as *Rattus rattus* was inadvertently introduced.

Within 5 years, the two of eh endemic murids, the Christmas Island rats (*Rattus macleari*) and the bulldog rats (*Rattus nativitatis*), were considered to be seriously affected. There was population declined in the precipitously, and the individuals are being described as behaving abnormally and even displaying evidences that focused on infection with the parasitic trypanosomiasis.

By the year 1908, it was observed that the native rats had become extinct, but there were some of the endemic diversity that should have survived in the form of *R. rattus* × *R. macleari* hybrids (which is deduced from the pelt characteristics).

Wyatt et al., often extracted the DNA from all the available remain Christmas Island and the bulldog rats along with the putative hybrid animals and black rats that are often collected at the same time. Nuclear, trypanosome, and mitochondrial DNAs were often amplified from the given samples.

Therefore, the mitochondrial and the nuclear DNA analysis clearly indicates that the *Rattus rattus* and *Rattus macleari* are biologically distinct. But the alleged hybrids are the extensively black rat morphotypes.

Therefore, the Christmas Island rat is found to be completely extinct and it is genetic endowment which has not persisted in any of the forms. Genetic evidences of the murid specific trypanosomes, *Trypanosoma lewisi* is often found in the true black rat and in the Christmas Island rat samples.

The samples of the bulldog rats that are collected earlier than the 1899 are considered to have no evidence of a trypanosome infection that can be found. Therefore, the detection of trypanosomes is found to be correlated with the arrival of an invasive black rats and subsequently the extinction of the native rat species on a Christmas Island.

Like other Island, the Christmas Island is considered to be notable for having suffered mammals' extinctions that cannot be easily explained by the hunting pressure or by the sudden changes in the climate.

The only other native ground that is found to be dwelling mammals on the island, the Christmas Island shrew (*Crocidura trichura*), which survived many decades after the disappearance of the endemic rodents.

But this Island has not been seen since 1985. Recently, the Christmas Island pipistrelle (*Pipistrellus murrayi*) often appeared to have become extinct in 2010. Which is followed by an unexplained precipitous decline which is starting in around the 1980s.

In Christmas Island, the last remaining native mammal was the Christmas Island flying fox (*Pteopus melanotus natalis*), which is also declining rapidly. These extinctions and declines are subsequently toward the loss of the endemic rats that are disease related and might be even resolved if there is an appropriate sample that was collected historically.

In regardless of the final cause, the rodents are considered extinctions on the Christmas Island which often display two interesting features. First, the extinction was extremely rapid which means that the interval from the first notice of decline is to declare the extinction which was 10 years, with the event that might actually be occurring within a shorter period.

Secondly, the species those are affected would not normally be regarded, as it is being particularly susceptible to extinction. Murid rodents are tending to extremely adaptable with the high reproductive rates. Therefore, they would be normally well buffered against the population which collapses sufficiently in order to cause the complete loss.

The susceptible factors which are at work in these cases are almost found to be certainly at a long isolation on and restriction to a single small island. Among the mammals, there is "island extinction effect" that is particularly well investigated.

There are more than 80% of the ~80 to 90 species of the mammals those might have become extinct in the past 500 years were the island is found to be dwelling. Such losses are usually found to be attributed towards the habitat destruction and the deleterious effects of invasive species, but they all are under some circumstances that often introduced the infectious disease that might be plausible as rivers.

There are various numbers of other, mostly anecdotal, examples of mammalian species that produces unknown infectious diseases. One extensive instance of this interest is the *Tasmanian tiger* or *thylacine* (which is also known as *Thylacinus cynocephalus*).

Never the less, it is extensively hunted because of its supposed taste for sheep, until the time period of the 20th century, these species (thylacines) were not uncommon. After that, thylacines became very uncommon, there were no confirmed kills recorded after the time period of 1910.

Distemper like outbreaks have been noticed in the captive thylacines, as well as extensive outbreaks in the other Australian marsupials at the same point of time, have been considered as anecdotal proofs that diseases played an act in the final diminishing of thylacines.

Verification of a role for disease in this extinction would be very challenging tasks, because distemper viruses are made up of single-stranded ribonucleic acid (RNA) paramyxoviruses which are not likely to be conserved in the museum skins or bones.

3.6.6. Overhunting

Overhunting is a process in which outcome as in critical decrease in the population of the species or harm to the wildlife. Overhunting is also defined as the relentless chase for wild or game animals with an objective of killing or catching the animals for economic as well as personal gains of for good.

In various regions of the world, human beings are continuing to rely on the wildlife for several numbers of various requirements and it is the act of killing, hunting down, or capturing more animals than their population can provide that contributes to the overhunting.

Due to the interlinkage of the ecosystem, the long- term impact of taking these kinds of species from their natural environment is overwhelming, and the reason behind this is that it not only impends their survival, but also, it threatens the survival of the other species as well.

From the time period of 17th century, more, and more hunting are logged in the form of the third most notorious cause of extinction of animals. According to the report of The WWF, which states that the overhunting is accountable for near about 23% of the extinctions. Several numbers of humans are forcing the animals into extinction by over-consuming them or killing the animals for their own economic profits.

In history, the activity of the overhunting is the main reason behind the extinctions of the giant lemurs and elephant birds in Madagascar, Miss Waldron's red colobus monkey in Ghana, moas in New Zealand, Alagoas curassow in northeastern Brazil and giant kangaroos in Australia.

Hunting is proven to interfere with the migration as well as the hibernation of the many animals. The reason behind this is that, the time, animals are being hunted or killed through the course of migration and hibernation, they develop fear of being hunted, which, at the end, makes these animals stop from hibernating or migrating.

The activity of overhunting not only impacts the animals, but also, overhunting also impacts the entire biosphere. As an outcome, it impacts the inter-association of the ecosystem, as the living organisms require each other together with their natural habit in order to survive.

In simple words, hunting disrupts the natural order and as each and every ecosystem is distinct and different from one another, the eradication of a living creature from their natural environment consequences as in an unbalanced ecosystem.

3.7. RECENT CASES OF INFECTIOUS DISEASE-CAUSING SERIOUS ENDANGERMENT IN MAMMALS

There are several numbers of various review articles have been analyzed the influence of the infectious disease on the existing natural population. With respect to this, there are several more instances may be added to verified association of disease and severe population decrement in the mammal species living in various environments.

3.7.1. Koalas

Koalas are under siege by two major pathogens. Kola retrovirus (which is represented as KoRV), every so often observed in captive koalas, is oncogenic and also, they cause significant mortality. The virus is not spread uniformly across Australia, as its frequency within the population reduced from north to south.

The history of the overhunting of the koalas in the last hundred years or more, more particularly in the region of southern Australia, it was originally determined that the KoRV must have penetrated Australia within the time interval of last two centuries, traveling from north to south.

Never the less, according to the latest study, which is utilizing the historical samples shows that the KoRV was already distributed by the time period of the late 1800s in the regions of northern Australia. In addition to this, its evolution is very steady, which is consistent with the retrovirus having entered the koala population long before this and to be expected that it is causing the disease all along this timer period.

At present interval of time, the second pathogen which is destroying the population of koala is a specific strain of infectious *Chlamydia pecorum*. Infection can outcomes in the form of blindness or sterility. There is some proof that Koala retrovirus infection is related to the chlamydial infection.

This is conceivable, as KoRV belongs to the family of gamma retrovirus, several numbers of the members of which are immunosuppressive. In this way, infection with KoRV would offer the chances for secondary infectious agents like Chlamydia is less- resistant hosts.

The population of koala shows low variations in their mitochondrial Deoxyribonucleic Acid (DNA). The low diversity of mitochondrial DNA produces lack of genetic diversity, if it impacts the overall resistance, may be emblematic of their inability to fend off infectious agents.

As per the reports of the World Wildlife Foundation, which states that serious overlapping infections of this sort could take into the direction of local extirpation, if not the outright extinction, of koalas within the time period of next five decades.

3.7.2. Myotis Bats

A cryophilous fungus (which is known as *Geomyces destructans*), introduced to the northeastern USA from Europe within the time period of last ten years, has been overwhelmingly shown to cause enormous mortality ("white nose syndrome") in the little brown bat (which is known as Myotis lucifugus).

Normally, fungal infections do not cause severe disease in mammals, apart from the cases of immunosuppression (which is every so often caused by another pathogen).

Fascinatingly, Myotis is closely correlated species endemic to Europe. And, it is also infected with this fungus, but other than the characteristic "white nose" produced with the help of fungal growth, animals suffer no negative symptoms and there is no increment in the mortality which is connected with the infection.

Geomyces destructans, which only cultivates at low temperatures, disturbs the hibernation of bats; this produces wintering animals to wake up more regularly as compared to their normal time, seriously impairing their energy and water balances and ultimately leading to death.

Even though, why these two species vary so vividly in their response to the infection is doubtful, the results are clear; hundreds of thousands of little brown bats, which are consisting of entire populations of some hibernacula, have died as an outcome of their incapacity in order to resist the infection.

3.7.3. Tasmanian Devils

Devil Facial Tumor Disease (DFTD) is a severe threat to the Tasmanian devil (which is known as *Sarcophilus harrisii*). At present interval of time, Tasmanian devil is receiving a great deal of attention because of its highly unusual nature.

Dissimilar from the most of the infectious diseases, in this case, the agent is not viral, bacterial, parasitic, or prion-derived (infectious protein).

In spite, the agent is a tumor which is spread from one body to the other body with the help of mechanical transmission of tumor cells through the course of agonistic encounters, which every so often involves aggressive behavior. At present interval of time, the only likewise infectious cancer recognized is known as canine transmissible venereal tumor (CTVT), which affects dogs.

The DFTD source tumor evidently derived from several numbers of individual clones that are developed in a single female; male and female devils are equally susceptible, and the disease has now spread all across Tasmania.

In the year 1996, the museum samples predating shows no proof of the disease, making it likely that DFTD is newly emergent. At the same point of time, there are complicated evolutionary dynamics at play; for instance, various kinds of clonal lineages that are derived from the original source tumor differ in frequency, geographical distribution, as well as the diversity.

How these kinds of tumors be able to escape the devil's immune system is unknown, but the ongoing threat is very real. Unless the tumor-free populations can be recognized and safeguarded, further population collapse, if not complete extinction, will be the probable outcome.

3.8. CONCLUSION

Wildlife has been declining at an alarming rate in recent years, which is also reflected in the various statistics that relate to the population of the wildlife. There have been numerous factors responsible for the decline and have collectively put enormous pressure on the wildlife resources.

The factors include various human activities such as hunting and poaching for leather or other materials or even trophy hunting. The human activities which include deforestation also put undue pressure on the wildlife to shift their natural habitat, which may also result in them losing their lives.

Other reasons for the loss of wildlife include the presence of uncontrolled invasive species, the wild animals getting infected with various kinds of diseases and the inability of the humans to treat them and the other environmental changes such as air and water pollution and climate change.

REFERENCES

1. Air-quality.org.uk. (n.d.). *Effects of Air Pollution and Acid Rain on Wildlife*. [online] Available at: http://www.air-quality.org.uk/17.php (accessed on 21 August 2020).
2. Brown, E., (2018). *Widely Misinterpreted Report Still Shows Catastrophic Animal Decline*. [online] Nationalgeographic.com. Available at: https://www.nationalgeographic.com/animals/2018/11/animal-decline-living-planet-report-conservation-news/ (accessed on 21 August 2020).
3. Conserve Energy Future, (n.d.). *Causes, Effects and Solutions of Overhunting*. [online] Available at: https://www.conserve-energy-future.com/causes-effects-solutions-overhunting.php (accessed on 21 August 2020).
4. Gooljar, J., (2018). *What You Need to Know About Declining Species*. [online] Earthday.org. Available at: https://www.earthday.org/populations-of-living-things-across-all-species-are-declining-and-this-is-very-worrisome/ (accessed on 21 August 2020).
5. Hundal, S., (n.d.). *Wildlife Conservation Strategies and Management in India: An Overview*. [ebook] Available at: https://www.arlis.org/docs/vol1/69415913/hundal_edited_final_march_10.pdf (accessed on 21 August 2020).
6. MacPhee, R., & Greenwood, A., (2013). Infectious disease, endangerment, and extinction. *International Journal of Evolutionary Biology*, [online] *2013*, 1–9. Available at: https://www.hindawi.com/journals/ijeb/2013/571939/ (accessed on 21 August 2020).
7. Mason, M., (n.d.). *Conservation: History and Future of Conservationism | Environmentalscience.Org*. [online] Environmentalscience.org. Available at: https://www.environmentalscience.org/conservation (accessed on 21 August 2020).
8. Nasi, R., & Brown, D., (2008). *Conservation and Use of Wildlife-Based Resources*. Montreal, Quebec, Canada: Secretariat of the Convention on Biological Diversity.
9. Naz, S., & Hussain, M., (2016). Current trends in wildlife conservation: A review. *International Journal of Fauna and Biological Studies*, [online] *3*(5), 44–48. Available at: https://www.researchgate.net/publication/308610297_Current_trends_in_wildlife_conservation_A_review (accessed on 21 August 2020).

10. Ndegwa, M. C., & Murayama, Y., (2009). Analysis of land use/cover changes and animal population dynamics in a wildlife sanctuary in East Africa. *Remote Sensing*, [online] *1*(4), 952–970. Available at: https://pdfs.semanticscholar.org/6fea/da8db37be9c57dee10674a27a24fe19d5946.pdf (accessed on 21 August 2020).
11. Olney, P., (2005). *Building a Future for Wildlife: The World Zoo and Aquarium Conservation Strategy*. [ebook] WAZA Executive Office. Available at: https://www.waza.org/wp-content/uploads/2019/03/wzacs-en.pdf (accessed on 21 August 2020).
12. Pacheco, P. X., (2019). How Technology Can Transform Wildlife Conservation. *Green Technologies to Improve the Environment on Earth*, [online] Available at: https://www.intechopen.com/books/green-technologies-to-improve-the-environment-on-earth/how-technology-can-transform-wildlife-conservation (accessed on 21 August 2020).
13. Quackenbush, C., (2018). *WWF Finds Human Activity is Decimating Wildlife Populations*. [online] Time. Available at: https://time.com/5438605/human-activity-wildlife-populations-wwf-report/ (accessed on 21 August 2020).

4

Population Assessment and Conservation Model Evaluation

CONTENTS

Population assessment is a very important task to be done in context to conservation and management of wildlife. This also aids the scientists and researchers to study the genetics of various species and create an appropriate environment for their survival.

This will reduce the chances of extinction of several endangered species. Effective conservation models and their evaluation are also being effectively used at present. In this chapter, the basic understanding of population monitoring is discussed along with certain measures related to it.

A detailed explanation of species and population monitoring (genetic and demographic monitoring) is given in the middle of the chapter. Towards the end of this chapter, a framework model for the evaluation of effective conservation and its basic features are discussed. Also, to explain better with an example a North American Model of wildlife conservation is given.

4.1. INTRODUCTION

Monitoring of populating is really an essential task as it will aid in effective management as well as conservation of wildlife (Nichols and Williams 2006). The information which is collected by monitoring of the population at different points in time can be effectively used to understand the dynamics of the population. It also detects the system-level responses to the management actions (Martin et al., 2007).

The capability of monitoring wildlife populations such as small endangered populations and elusive species, has expanded much after the use of Deoxyribonucleic acid (DNA) and population genetic data that allows researchers to investigate genetic as well as demographic parameters and also, to enumerate their temporal changes that is genetic monitoring (Schwartz et al., 2007).

The use of DNA-based monitoring programs by non-invasive genetic sampling (NGS) is becoming highly recognized in the community of wildlife (Waits and Paetkau 2005; Long et al., 2008). although, a very small number of studies have applied NGS for the detailed genetic monitoring over several years.

Long-term monitoring is specifically important after the process of translocation or reintroductions in order to evaluate the real success of all these kinds of programs and to be assured that quick actions are being taken for the improvement of the status and the probability of population persistence (Miller et al., 1999).

Demographic information like recruitment, reproduction, age structure, sex ratio and mortality need to be assessed at certain time intervals that are biologically relevant to evaluate the dynamic rates and their variances, trends in population and the impact of environmental and demographic stochasticity (Sarrazin and Barbault, 1996; Miller et al., 1999; Sed-don, 1999).

Understanding the ecology of introduced species, along with the patterns of geographical expansion from the land are of reintroduction, connectivity with neighboring populations, dispersal behavior, identification of potential sink habitats and habitat use, is important for the process of conservation as well as management (Miller et al., 1999; Frankham et al., 2002).

In small populations, the mating behavior is impacted by the presence of very small number of mates. Hence, knowledge, and understanding of mating together with social structure is critical for monitoring effective sizes, genetic structures, and also, the rate of inbreeding accumulation (Chesser, 1991; Sugg et al., 1996, 2005; Vonholdt et al., 2008).

In general, the reintroduction programs release a small portion of animals and thus, the effective size of the population is small at first. As a result, changes in monitoring in genetic parameters is of key concern for the viability of the population (Lande and Barrowclough, 1987; Frankham, 2005).

In the absence of gene flow, certain levels of coancestry are expected to increase; as a result of drift and mating between relative's genetic diversity is lost to a great extent; the adaptive evolutionary potential is limited and the growth of deleterious alleles is accelerated (Ralls et al., 1988; Hedrick, 2000).

In spite of their identified significance in the phases following the process of reintroduction, very detailed monitoring programs are usually overlooked and their results are not reported (Sarrazin and Barbault, 1996; Ostermann et al., 2001), and genetic monitoring is underused at present (Schwartz et al., 2007).

A recent translocation of brown bears (Ursus arctos) in northern Italy (Trentino) provides a unique case study for the use of genetic monitoring for small and reintroduced populations. Originally, Brown Bears were present all around the European continent, though during the 20th century, they were not present in the south-western Europe, they got eradicated (Swenson et al., 2000).

The decrease in the population of bear in northern Italy started around the 18th century and it was continued till the 20th century due to human persecution, fragmentation, and loss of habitat (Mustoni et al., 2003).

In the Italian Alps, bears became isolated from the population that once upon a time extended from the Balkans to the Alps. By the year 1990, just a few relict individuals survived in Trentino and therefore, the population was regarded as functionally extinct (Mustoni et al., 2003).

A translocation plan was initiated in the year 1999, for the re-establishment of a viable bear population in the central Alps (almost 50–90 bears) (Mustoni et al., 2003). between year 1999 and 2002, a total of nine bears captured in Slovenia were released out in Trentino. All the released bears were radio monitored during the initial phase. After the loss of radio-collars, genetic monitoring was the most preferred method over traditional field-based methods for tracking the relict and re-introduced bears.

4.2. MONITORING OF WILDLIFE POPULATIONS

The population of wild animals can be monitored for several reasons (Witmer, 2005) such as:

1. Biological interest or research purposes, for example, bird migration;
2. Game management for example for deer;
3. The wild animal may be considered an actual or potential pest like birds and rodents;
4. The species may be threatened or endangered;
5. The population is monitored for assessing its recovery or progress;
6. For checking the status of biological diversity and ecological health of a specific area; and
7. It might be desirous to know the impact of human management actions, land-use practices, or alternative activities on one or more species.

In order to monitor the population of wild animals, first of all information related to the animal interest need to be collected. This is however achieved by using certain methods like:

1. By observing animals and their behavior in situ directly;
2. By looking for certain signs of animals such as dung, tracks, etc;

3. By using radar;
4. By using thermal cameras;
5. Just by capturing, marking, and release of animals; and
6. Implanting or attaching monitoring devices like leg bands, collars, data loggers or backpacks.

Based on the kind of method used for the collection of data, the impact on the animal's welfare may be less or more, both for the individual which is being monitored and the population as a whole. For instance, to make field biologists enable in monitoring wild penguin populations, flipper bands can be utilized to mark individual penguins in a penguin colony.

Nevertheless, this practice is really critical for the collection of population data, the welfare of banded penguins can be compromised at the time when traditional metal bands are used. They may eventually contribute in the reduction of population over a certain period of time.

These metal bands cause increase in drag at the time of swimming. This in turn, decreases the speed of swimming along with the dive depth, it also impacts feeding. It can snag on vegetation and cause injury.

Ultimately it may result in increased feather wear which leads to an increase in heat loss (an important consideration for birds that live in Antarctic conditions). It is important to note that banded penguins have almost 16% high rate of mortality and produce around 39% of less chicks (Saraux et al., 2011).

When a method of identification and monitoring affects the welfare of animals detrimentally, it is essential to consider some alternatives. In recognition of Dr. Peter Barham's and Bristol Zoo's work on the welfare benefits of a new kind of silicon flipper band that addressed several concerns of metal bands UFAW awarded them a "UFAW Wild Animal Welfare Award."

The silicon flipper band that was designed by Dr. Barham of Bristol University, manufactured, and tested by Bristol Zoo, was a very simple as well as innovative design. This was found to cause quite less feather wear and drag. It was easy to read. In addition to that, the flexible nature of the rubber allowed the bands to expand if required, for example, at the time when the flipper expands during moult.

Another project that was initiated for improving the welfare of wild animals fitted with the external tags is supported by UFAW. This project is being carried out by Rory Wilson and his team; they are investigating tag-use in free-ranging birds. Guidelines, at present, state that the devices should not exceed 3% of the body mass of birds. The key aim was that the device does not interfere much with the behavior and welfare of birds.

Although, according to Rory Wilson and his team, this measurement is very simplistic and does not consider bird biology as well as lifestyle. The objective of this project is to create a framework for evaluating assumed physical tag detriment on birds, for examples, skin pressure, differences in predicted metabolic rate, drag, moment arm effects, etc. using physical principles that could be used once the tag properties and the study species were known.

A generated 'physical detriment matrix' as a result could be used to give a realistic measurement of the detrimental tag effects. As well, the internal devices can be essential for the monitoring of animals. UFAW supported Mr. James Macgregor's work investigating platypus monitoring techniques (Murdoch University, Australia). Wild platypuses in Australia are monitored in order to assure their welfare and health, and also to protect the population.

Although long-term monitoring methods consist of either repeated capture or the use of comparatively large tracking devices, both of these can affect the individual platypus welfare in a very adverse way. Mr. Macgregor tried to refine techniques of monitoring by using an in-stream microchip platypus movements and remotely survival and needed only one capture of each platypus in order to implant the microchip.

The study showed that in-stream microchips were a significant method for the detection of micro chipped platypus and Dr. Macgregor believes that in-stream microchip readers will greatly support in collecting new data as well as improving platypus welfare. The complete results of Macgregor's work were published in the Journal: Pacific Conservation Biology.

4.3. SPECIES AND POPULATION MONITORING

4.3.1. What Is It?

Species and population monitoring are a kind of regular observation and recording of certain changes in status and species trend or their population in a specific territory. The key aim of such monitoring is to gather information that can be utilized in examining the results of management actions and to guide the decisions of management.

This is often been carried out for species that have been evaluated as a threat in order to determine when actions of conservation are important or when the current ones required to be intensified. In the case of CWR, it may be important to monitor the population size, numbers, structure, density, and demographic variables as a key part of the evaluation of their conservation status, and the following impacts on the CWR populations of management interventions that are agreed in the management plan of species so as to judge their effectiveness.

Species and population monitoring programs such as biological monitoring is remarkably variable as well as diverse in large scale, aims, and coverage. In the year 2008, Marshand Trenham tried to detect the trends in plant and animal population monitoring.

The aims and strategies exhibited signs of diversifying with some of the approaches that are becoming very frequent like occupied areas and presence or absence approaches. On the other hand, others have yet to be applied widely for example, risk-based monitoring and linking the outcomes of monitoring straight to management decisions.

Therefore, it is important that the aims and objectives of any proposed population monitoring are defined clearly in advance (Yoccoz et al., 2001). A lot many samplings as well as analytical techniques are present in the world for the monitoring of species and population and are reviewed by Stork and Samways (1995) and specifically for CWR by Iriondo et al. (2008).

Key Steps involved in a Monitoring Program are:

1. Complete background tasks;
2. Develop objectives;
3. Design and implement management;
4. Design monitoring methodology;
5. Implement monitoring as a pilot study;
6. Implement and complete monitoring; and
7. Report and use results.

4.3.2. Which Attributes to Monitor?

The attributes of species for which the goal and targets of monitoring may be set consist of (Tucker et al., 2005):

a. Range;

b. Abundance;

c. Demography;

d. Habitat requirements;

e. Population dynamics (recruitment, emigration, mortality); and

f. Population structure (age, sex ratio, genetic diversity)

4.3.3. Demographic Monitoring

Demographic monitoring is a very common form of monitoring population mainly for rare or endangered species. This will usually be found to be a suitable method or approach for CWR where the main focus is on the maintenance of populations that are viable along with their genetic variability.

However, demographic monitoring is an evaluation of changes that take place in population together with their causes all over the life cycle and measures the attributes or characteristics such as mortality and germination rates, size, density, growth, and distribution. It can also be utilized in aiding the establishment of factors that determine the abundance as well as the distribution of species and forecast the future structure of populations.

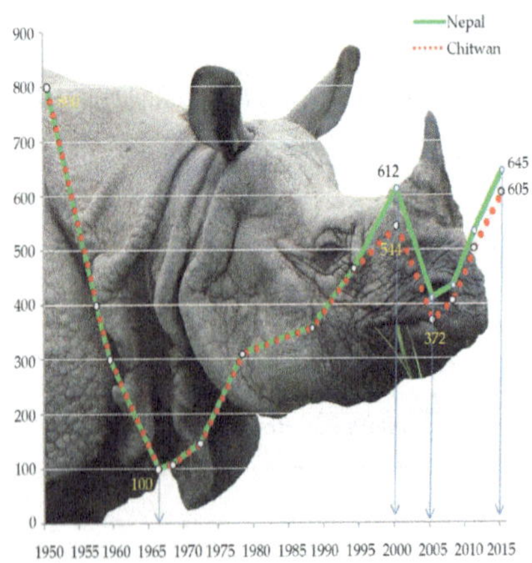

Figure 4.1. A depiction of the wildlife population assessment.

Source: Image by Wikipedia.

This demographic monitoring may involve regular measurements or mapping if an important level of resolution is to be attained (Given, 1994). There are some major demographic approaches, such as:

i. Analysis and availability of population;
ii. Single age or stage class investigations; and
iii. Demographic structure (Elzinga et al., 1998).

Demographic approaches in monitoring are usually quite time consuming and expensive. Therefore, these are not feasible always. In addition, a demographic approach may not be suitable for some specific situations (Elzinga et al., 1998).

For instance, caution against the improper use of demographic monitoring for particular kinds of species, especially those with dense vegetative reproduction, long-lived seed banks, episodic reproduction, very short or very long lifespan, multiple stems and mat-like morphology, large populations and high densities in heterogeneous habitats (Elzinga et al., 1998).

4.3.4. Genetic Monitoring: What Is It and When to Use It?

The present era of rapid change in global environmental change (Zalasiewicz, Williams, Haywood and Ellis, 2011) is projected to lead to a rapid loss of biodiversity (Pimm et al., 2014). To evaluate and mitigate the impact of such loss, a lot many national, as well as international organizations have established strategies of biodiversity monitoring (Kurtz, Jackson and Fisher, 2001; United Nations Environment Programme Convention on Biological Diversity SBSTTA, 2003).

Key tools for biodiversity monitoring make use of methodological approaches from the field of genetic monitoring, based on genetic tools for evaluation of change (Stetz, Kendall, Vojta and GeM, 2011). Genetic monitoring focuses completely on quantifying the temporal changes genetic matrices of population, or other population data, produced using molecular markers (Schwartz, Luikart and Waples, 2007).

Genetic monitoring can be utilized to estimate various biological parameters of interest, comprising demographic parameters like vital rates, abundance, occupancy, disease status, hybridization; population genetic parameters such as genetic diversity, effective structure and population size; and gradually, responses to selective pressures like exploitation such as trophy hunting and climate change (Schwartz et al., 2007; Stetz et al., 2011).

Approaches for genetic monitoring that use non-invasive for example, naturally shed feathers or minimally invasive samples, for example, buccal swabs, biopsy darts are examined because wildlife ecology and its conservation has greatly benefited from the new data provided by these kinds of approaches (Beja-Pereira, Oliveira, Alves, Schwartz and Luikart, 2009).

Genetic monitoring by using MIS approaches were initially introduced in the year 1992 as an approach for the collection of genetic samples from brown bears (Ursus arctos; Höss, Kohn, Pääbo, Knauer &Schröder, 1992; Taberl et and Bouvet, 1992) and to study social structure in chimpanzees (Pan troglodytes; Morin and Woodruff, 1992).

MIS has now become a method of choice for genetic monitoring of various vertebrate species. This is mainly due to sampling of hair, faces, feathers or remote skin biopsies provides DNA from free-ranging animals that can be utilized to recognize individuals across space and time. It also produces genetic data without having to catch, handle or even observe them in certain cases (Beja-Pereira et al., 2009; Schwartz et al., 2007; Waits and Paetkau, 2005).

In the past 25 years, researchers have demonstrated a broad range of significant applications of MIS such as:

a. Detection of rare species (Palomares, Godoy, Piriz, O'Brien and Johnson, 2002; Valière et al., 2003);
b. Estimation of population size and certain other demographic parameters (Carroll et al., 2013; Kendall et al., 2009; Kohn et al., 1999; Rudnick, Katzner, Bragin, Rhodes and DeWoody, 2005; Woodruff, Lukacs, Christianson and Waits, 2016; Woods et al., 1999);
c. Evaluation of genetic diversity as well as gene flow (Epps et al., 2005; Gerloff, Hartung, Fruth, Hohmann and Tautz, 1999; Lucchini et al., 2002; Palsbøll et al., 1997);
d. Detection of migration and movement (Dixon et al., 2006; Proctor, Mclellan, Strobeck and Barclay, 2005);
e. Evaluation of social structure (Constable, Ashley, Goodall and Pusey, 2001; Ford et al., 2011; Morin et al., 1994);
f. Detection of hybridization (Adams, Kelly and Waits, 2003; Bohling et al., 2016; Steyer et al., 2016);
g. Monitoring disease epizootics (Kohn and Wayne, 1997; Schunck, Kraft and Truyen, 1995);

h. Identification of diet items (De Barba et al., 2014; Höss et al., 1992; Taberlet and Fumagalli, 1996); and

i. Other wildlife applications (Banks, Horsup, Wilton and Taylor, 2003; Ernest, Rubin and Boyce, 2002; Lukoshek et al., 2009; Wasser et al., 2004).

Now, there are huge amount of published evidence that MIS is easily comparable in costs or much more cost-effective (De Barba et al., 2010; Solberg et al., 2006) as compared to traditional methods (e.g., camera trapping, tracks, and signs and even trapping animals).

It is also evident that collection, as well as analysis of larger genetic sample sizes are usually possible (Garton, et al., 2010; Marucco et al., 2009; Solberg et al., 2006; Stenglein, Waits, Ausband, Zager and Mack, 2010). This eventually prompts a lot many wildlife managers to shift to MIS approaches.

A major methodological and analytical development has been invested in the establishment of protocols to increase the rates of success and decrease the rates of error while using such low-quality DNA sources for genetic monitoring (Beja-Pereira et al., 2009; Broquet and Petit, 2004; Miquel et al., 2006; Morin et al., 2010; Smith and Wang, 2014; Taberlet et al., 1996; Taberlet and Luikart, 1999; Waits and Paetkau 2005; Wang, 2016).

Genetic monitoring is all set to become very essential task in the future, mainly because a lot many vertebrate species have undergone a rapid decline in anthropogenic population (Li et al., 2016) that are doubtful to decline without severe management efforts.

Luckily, the genetic revolution of the early 2000s has eventually given rise to a broad range of more powerful and more precise molecular techniques that will make the process of genetic monitoring much more effective in the coming era. Novel and modern technologies for genetic monitoring depends largely upon single nucleotide polymorphisms, or SNPs (Morin Luikart and Wayne, 2004).

Different from more traditional DNA markers like SNPs and microsatellites have comparatively very less alleles per locus (in theory, up to four but in general, just two because of low mutation rates) (Glaubitz, Rhodes and DeWoody, 2003) and generally have more limited use across a variety of species as compared to microsatellite markers, frequently being species-specific.

In addition to that, SNP loci are very prone to the ascertainment bias, as they are mainly selected due to their polymorphism in the population of interest. However, they are often monomorphic in even closely related populations (Gautier et al., 2009).

Although, approaches based on SNP have major potential for non-invasive genotyping as:

1. High numbers of loci can be surveyed at the same time, mainly with the next-generation sequencing or genotyping platforms.
2. A comparative ease of scoring, modeling, and analysis of SNP genotype data because of the digital or binary nature of the data. The final point favorably contrasts with the continuous distribution of microsatellite alleles which can be quite tough to characterize and hence, could cause certain scoring errors.

4.4. A FRAMEWORK FOR EVALUATING THE EFFECTIVENESS OF CONSERVATION ATTENTION AT THE SPECIES LEVEL

In the current years, there has been a rise in efforts to determine the efficiency of conservation interventions (Christensen, 2003; TNC, 2007; Kapos et al., 2008; Black and Groombridge, 2010). It is important to understand whether the interventions of conservation are having the wanted effect, mainly in the light of rapidly increasing pressures on environment or biodiversity (Butchart et al., 2010) and the increased concern of donors that the projects they are supporting are working (Redford and Taber, 2000).

Most of the proposed methods focus specifically on calculating the effectiveness at either a project level (O'Neill, 2007; Kapos et al., 2008) or at the organizational level (Christensen, 2003). At the same time as evaluating the effectiveness at these levels is essential, there is also a requirement to connect local achievements with the global distribution as well as the status of a species to understand the efficiency of conservation directed towards a species as a whole (Saterson et al., 2004; Kondolf et al., 2008; Redford et al., 2011).

However, there is a shift to ecosystem-based approaches (MEA, 2005), most of the metrics regarding biodiversity, such as measurement of conservation success, are expressed in units of species (Mace, 2004; Garnett and Christidis, 2007; Tobias et al., 2010).

Thus, there is a mismatch between the project-level and organizational-level evaluation tools available for the conservationists and the species focus of various global conservation programs. Conceptual models for the visualization of components of a conservation intervention includes the stages of input, output, and outcomes along with two additional stages, strategy, and impact. These additional stages are added to the beginning and end of the list respectively (Woodhill, 2000; Ferraro and Pattanayak, 2006; Margoluis et al., 2009).

Input denotes to the resources that are committed to an intervention or certain plans for the action (Woodhill, 2000). While outputs, in general, comprise tangible products and often they are reported as a measure of success and can comprise awareness material produced, money spent or people trained (Mace et al., 2007).

Although, the presence of outputs does not assure they have had their desired effect. The outcome clearly considers the impact that the outputs have had on the state of the targeted species. Well, a better measure of success, outcomes are reported very rarely as they are usually intangible and hence, quite difficult to measure, or take place over a longer time-scales as compared to standard funding or reporting cycles (Kapos et al., 2008).

The conservation interventions are led by the conservation attention, that is, attention should be drawn to the issues and difficulties of a species. The plans and preparations can be made which will be carried out before the implementation of interventions. In the year 2009, Sitas et al, proposed 'an index of conservation attention' that makes use of a very simple and ordinary scale to rate the attention related to conservation, being directed at a specific species. It is basically measured by the presence as well as the quality of a species action plan.

The first major attempt is to understand the efforts of conservation at a large scale across whole sets of species underlines the significant idea of conservation attention. Although, the presence of a plan does not assure its actions will surely be carried out of the actions will eventually lead to an effective conservation (Kapos et al., 2009).

The final goal of conservation can take various decades to be attained (Redford and Taber, 2000). It can be quite discouraging to work towards a definite goal that is unlikely to be attained in the career span of an individual (Mace et al., 2007), and where the projects mention a projected time-scale it is rarely longer than just a few decades (Redford et al., 2003).

The Nature Conservancy's Site Consolidation Scorecard reports this fairly by developing a framework to measure the success of a protected area against predetermined criteria which once met are presumed to prepare the protected area with the ability to conserve the biodiversity within its borders successfully (TNC, 2007; Leverington et al., 2008).

Although, this scorecard approach has not been utilized for the conservation of species. Therefore, it would be of great use to have an index of the significance of conservation attention that measures the progress over a certain period of time from efficient conservation inputs to outputs and outcomes. This can further be used to observe the key impact of conservation actions for a particular kind of species and offer an early warning system if the conservation attention has stuck.

One main reason why the effectiveness of conservation attention is highly tough to evaluate is that most of the information related to conservation of species is not yet published, or is not accessible in an easy way, at times, even to working in the similar species (Brooks et al., 2009).

Where the data are published, the effective time lag between collection of data and its publication can mean the condition for a species has changed already from that reported in a publication (Leverington et al., 2008).

Several experts involved in the conservation of species are an intact resource of updated species information (Hockings, 2003) and can usually draw on their knowledge of published as well as unpublished data to give a viewpoint on projected future developments, something that cannot be attained by a non-expert reviewing the literature (Johnson and Gillingham, 2004).

Nevertheless, there are some uncertainties with expert elicitation that comprise but are not restricted to:

1. An inclination towards overconfidence in estimates (O'Neill et al., 2008);
2. Anchoring the estimations to a provided or pre-conceived range or value, and consequent inability to adjust the value effectively in context to the anchor (McBride et al., 2012);
3. Discrepancies with the terminology (Johnson and Gillingham, 2004);
4. Control of information though an inability to show it represents a poor design of survey (Martin et al., 2012) or framing of information in a different way according to presentation (McBride et al., 2012);

5. Confirmation bias in which the answers are probably interpreted in the regard of pre-existing beliefs of experts (McBride et al., 2012); and
6. Accessibility bias, where certain pieces of information are easily recalled as compared to others (Martin et al., 2012)

The Evolutionarily Distinct and Globally Endangered (EDGE) index line up species for conservation by applying a combination of the extent of the threat that they are facing and the volume of evolutionary uniqueness that would be lost were the species to go extinct (Isaac et al., 2007).

The EDGE program at the Zoological Society of London (ZSL) mainly concentrates on the conservation of the top 100 EDGE ranked amphibians, mammals, corals, and birds (EDGE program, 2013a). The significant goals of the EDGE program comprise increasing awareness of the concerned species, and tracking the improvement of their conservation (EDGE program, 2013b).

Figure 4.2. EDGE foundation is working extremely hard to conserve wildlife.

Source: Image by Wikipedia.

ZSL recommends to track the improvement by making use of species report cards that actually summarize different aspects of the conservation of a distinct group of species (Sinfield, 2011). These report cards are completely based on the format pressure-state-response, and it can be used for understanding the threats along with their impacts (Zalidis et al., 2004; Mace and Baillie, 2007; Roura-Pascual et al., 2009).

As the EDGE program mobilizes a big group of experts and as it is concentrated on the improvement of the conservation status of species as a whole, it is a potential case study for recognizing the latent for a species-based index of conservation attention that associates the easiness of a species report card approach with the information on the progress of conservation action within a definite range of species, making practitioners able to track the attainment of milestones.

A novel framework for the assessment of the significance of conservation attention that is directed at a particular species across its whole range is important to have in action. The framework that takes into account the on-the-ground interventions concerning a species instead of concentrating on a specific organization, connects action at a local scale with the change in the world-wide status of a species.

It is however applied for use by the experts of definite species such as the members of the IUCN Species Survival Commission specialist groups, as they can easily monitor the success of conservation attention over time.

> According to Leptz et al. (2009), it is quite tough to forecast the long-term biological responses as people have very less knowledge regarding lag-times between a given effect and its response. To represent and understand the change in climate and its impacts on biodiversity, it is important to monitor population of species over a very long period of time, as effects are only detectable after various years.

4.5. UTILITY OF POPULATION MODELS

Models of population dynamics that are also referred to as population models are found to be very essential tools that aid in understanding, explaining, and forecasting the dynamics as well as the determination of the biological populations.

From the management point of view, such kinds of models can be utilized for the assessment of the status of a specific population. It also diagnoses the cause of decrease in population along with rapid increase. It has been observed that the population models are applicable in prescribing management targets and assessing the prognosis of a population's responses to alternate management actions (Caswell, 2001).

For example, United States (Crouse et al., 1987; Crowder et al., 1994; Caswell, 2001) used population dynamics modeling as an important tool in reversing the decrease of endangered loggerhead sea turtle population.

Till the year 1980s, efforts regarding the protection of sea turtle had mainly focused on the security of nets eggs and hatchlings on nesting beaches. After the study of stage-structured population models, it was discovered that the sea turtle's growth rate was proportionately more complex to change in survival and that decreasing the mortality of adult and sub-adult turtles at sea would be a most effective method of growing population increase rate as compared to defending nests and hatchlings on nesting beaches.

As informed in part by all those findings, certain rules and regulations were forced to require turtle excluder devices (TEDs) in shrimp trawls in the range of sea turtle. However, it was initially controversial, the application of TEDs was endorsed by a National Research Council Committee (NRC, 1990). In addition, it is thought to have had an effective favorable impact on the populations of the loggerhead sea turtle (Caswell, 2001).

Population dynamic model can also useful to forecast the population's reaction to environmental changes for example, global climate change. Global climate change is forecasted to impact the Arctic sea in a very adverse manner, and population dynamics could also be affected by this along with the persistence of species that depends on sea ice environment.

For example, feeding as well as breeding of polar bears totally depends on arctic sea ice. After integrating the field analysis data, climate change models, and population model, Hunter et al. (2010) forecasted that the polar bear population in the southern Beaufort Sea would experience a strong decline because of a decrease in sea ice extent by the end of the 21st century.

The population models are found to be very effective tool that can be used for the management of abundant species. For instance, the American bullfrog is an introduced species on Vancouver Island which is affecting the biodiversity on parts of the island very badly.

A modeling study conducted by Govindarajulu et al. (2005) suggested that the management strategy of targeting the eradication of tadpoles may not be very successful because incomplete removal of tadpoles could ultimately lead to higher tadpole survival and in turn, this leads to owing reduced density-dependent effects.

Their results showed that removing metamorphs in fall would be very efficient in controlling the population of bullfrogs. A theoretical study done by Zipkin et al. (2009)recommended that the control of excessive species by harvest or removal could backfire. This is mainly because the populations of species categorized by early maturity as well as high fecundity may experience rapid increase after harvest or removal as a consequence of density-dependent overcompensation.

Certain other examples of the use of population models comprise:

1. The assessment of the impact of culling and fertility control on the population dynamics of an excessive population (Bradford and Hobbs, 2008);
2. Controlling the fertility of the \koala on koala-forest dynamics (Todd et al., 2008)
3. Evaluation of the efficiency of euthanasia versus trap-neuter-return for the critical management of free-roaming cats (Andersen et al., 2004);
4. The effectiveness of fertility control in the population of a white-tailed deer (Merrill et al., 2003);
5. Discerning mechanisms that underlie a rapid increase in current population in case of yellow-bellied marmots (Ozgul et al., 2010);
6. Forecasting the impact of El Niño on the dynamics as well as determination of the Galapagos penguin population (Vargas et al., 2007);
7. Evaluating the impact of harvest on the persistence of dugongs (Heinsohn et al., 2004); and
8. Predicting the effect of estimated climate change on the dynamics as well as the persistence of emperor penguin populations (Jenouvrier et al., 2009).

4.5.1. Basic Features

At the very basic level, a good population model would represent an accurate free-ranging life-history, mating system and social structure. It would also consist of certain factors along with some processes that can really affect the dynamics of the population such as demographic and environmental variability, and density dependence.

Climatic variability can significantly impact the growth of the population through its key effects on the availability of forage and as a result, survival, and reproduction. Clear connections between data related to weather and demographic vital rates would increase the realism of virtual scenarios.

Figure 4.3. Climatic variability has a great role to play in the population assessment of wildlife.

Source: Image by Columbia University.

It has been found that a lot many vital demographic rates can be influences in a potential manner by an increase in competition for forage at very high population densities, mainly if the populations are permitted to raise the food-limited carrying capacity.

Hence, options to allow vital rates to be density-dependent, and possibly the presence of alternative functional forms for the density-dependent effects, might be very useful. However, there are few studies of mechanisms that produce density-dependent responses in free-ranging burro and horse populations.

The data-based estimates of some parameters that describe the relationships between climatic variables, population density, and demographic vital rates were not present at the time of study.

Transient and asymptotic sensitivity analyses (that is, the sensitivity of asymptotic growth rate of population, expected size of population, or chances of extinction to vital demographic rates along with some other input parameters) are really essential tools that have been utilized in setting up the priorities for research and making the decisions related to the management of population (Crowder et al., 1994; Caswell, 2001, 2005, 2007).

4.6. THE NORTH AMERICAN MODEL OF WILDLIFE CONSERVATION

The North American Model of Wildlife Conservation is a complete set of principles that is applied collectively and thus, led to the form, function, and successes of wildlife conservation as well as management in the United States and Canada.

This technical review documents the history along with the development of those principles and assesses potential future challenges to their effective applications. Defining the model as North American is done in a very conceptual context, not in a geographical context.

Conservation and management of wildlife in Mexico developed at a different time and under distinct conditions than in Canada and the US. The latter two were hand in hand. The history, development, and status of the wildlife conservation, as well as management in Mexico, are separately highlighted.

Figure 4.4. Wildlife is a very significant part of the ecosystem and deserves best measures to be managed.

Source: Image by Needpix.

It is not the purpose or intention of this review to revise, alter, and modify what has been put forward as the model. However, the model itself is not a monolith that is carved in stone. It is a means for people to understand, assess, and celebrate how the process of conservation has been attained in the US and Canada, and also to evaluate whether people are prepared to address key challenges laying ahead.

Just adding to, deleting or altering the current principles will not just advance the process of conservation. Understanding the evidentiary basis for the principles is important for the prevention of erosion and also for the conceptual thinking needed for anticipating future challenges.

A brief summary of some key challenges and associated concerns are given below:

Wildlife resources are a public trust: The challenges consist of-

- Inapt claims of ownership of wildlife;
- Unregulated commercial sale of live wildlife;
- Restrictions or prohibitions on access and use of wildlife; and
- A kind of value system that endorses an animal-rights doctrine and also, antithetical to the premise of public ownership of wildlife.

Markets for the Game are Eliminated: Commercial trade exists for fish, amphibians, and reptiles. Additionally, some of the game species are traded in an active manner. A healthy market for access to wildlife across a country occurs in the form of reserved permits, leases, and shooting preserves.

Allocation of wildlife is Done by Law: Execution and enforcement of laws to all taxa are not consistent. However, state authority over the allotment of the take of resident game species is very well defined, local, county, and housing development ordinances may highly supersede state authority. Decisions on the use of land even on the public lands impact the allocation of wildlife indirectly because of changes in use if land associated with the development of land.

Wildlife Can Be Killed for a Valid Reason: take of some species of birds, mammals, amphibians, and reptiles does not correspond to traditionally accepted concept of legitimate use.

Wildlife is Regarded as a Global Resource: various positive agreements as well as cooperative efforts have been established amongst

the US, Mexico, Canada, and several other countries for the conservation of wildlife.

A Lot Many Species Require Consideration: restrictive procedures for permitting, however, designed to protect and conserve the wildlife resources and inhibit trans-border collaborations. Making of a wall in order to prevent illegal immigration to the US from Mexico will have negative effects on trans-border wildlife movements and their interactions.

Science is an APT Tool to Discharge the Policy of Wildlife: the management of wildlife appears to be politicized gradually. The rapid rate of turnover of state agency directors, the makeup of boards and commissions, the organizational structure of certain agencies, and instances of political meddling in the field of science have challenged the foundation of science.

Democracy of Hunting is Standard: Decrease in and access to huntable lands eventually compromise the belief of egalitarianism in the opportunity of hunting. Restrictive firearms legislation can act as a barrier and obstructs the participation.

4.6.1. Historical Perspective

The exploration of whole North America by English and French was principally motivated by the wealth of natural and renewable resources of the continent and an unconstrained opportunity by the individuals in order to exploit them (Cowan, 1995).

At present, wildlife conservation in the US and Canada reflects historic citizen access to the land and its natural resources. Certainly, the sense that all these resources have their places in the citizenry and drives the democratic involvement in the process of conservation and it is the raison d'etre of North America's unique approach (Krausman, P., Gold, silver, and souls, unpublished presentation at The Wildlife Society Annual Conference, 22 September 2009, Monterey, CA, USA).

Exploitation of resources fueled the expansion of people all across the continent and led to the disappearance of the frontier eventually (Turner, 1935). The Industrial Revolution brought key changes to the society of North America that changed the land as well as its wildlife.

In the year 1820, 5% of Americans lived in cities or urban areas, while by the year 1860, 20% were urban dwellers, a 4-fold increase that marks the major demographic shift ever to have taken place in America (Riess,

1995). Markets for wildlife ascended to feed all urban masses and drape a completely new class of all the wealthy cities.

The hunters of the market first practiced their trade along the coastal waters as well as interior forests. With the development of railways and refrigeration, they exploited elk (*Cervus elaphus*), bison (*Bison bison*) and various other big games of western North America for the transport back to different cities in eastern North America.

All market hunters left various species teetering on the edge of extinction, that were once present in very abundant amount. Unluckily, the absolute scale of such unmitigated exploitation was to have some major impacts on endangering as a remarkable phenomenon: protectionism and conservation (Mahoney, 2007).

Meanwhile, the increase in urban population found themselves with certain things that their countrymen on the farms did not have and that thing is called as leisure time. The act of hunting for the rigors, and the issues that come across while chasing under certain conditions of fair play eventually became a favored time pass activity of a lot of many people, especially amongst those of means.

This developed in situ. However, there is no doubt that European aristocratic viewpoints toward hunting exerted few impacts on the emerging trends (Herbert, 1849). Threlfall (1995), noted that European commoners never left the desire to get actively involved in the hunt, in spite of this, the best and brutal efforts of nobility was made in order to discourage them.

In the US, after some time, conflicts arose between the market hunters who used to get profit in dead wildlife and the new breed of hunters who actually placed value on live wildlife as well as their sporting hunt of it. Such sport hunters organized and then developed the first refuges for wildlife (Carroll's Island Club, 1832; Gunpowder River in Maryland; Trefethen, 1975) along with laws to protect the game (New York Sportsmen's Club, 1844; Trefethen, 1975).

Representatives of these sport hunters was George Bird Grinnell, a highly influential and a yale-educated naturalist who accompanied George Armstrong Custer on his Black Hills excursion and who attained the sporting journal Forest and Stream in the year 1879.

Over the next thirty years, George Bird Grinnell would turn 'Forest and Stream' into a 'call for the conservation of wildlife.' Later, in the year 1885, he reviewed a book written by a fellow New Yorker related to his hunting exploits in the Dakotas (Grinnell, 1885).

His review was admiring, though he criticized the author for certain inaccuracies. Theodore Roosevelt, the author, went to meet George Bird Grinnell and then both of them realized that a lot had changed in the past one decade that ultimately divided their respective times in the West and that big game animals had drastically declined.

The discussion what they did together inspired them to establish the Boone and Crockett Club in the year 1887. It was an organization whose key purpose was to take key charge of all the matters or cases related to the enactment and carrying out of game as well as fish laws (Reiger, 1975:234).

Both of them were also known as nation builders who felt that America was a very strong country because, similar to Canada, people of America had carved the nation out of a wilderness edge with pioneer skills and self-reliance. This, in turn, harkened back to the ideals concerning the impacts of the frontier on making what it is to be an American ideals were expressed in the late 19th century by Turner (1935).

Turner described the very romantic concept of primitivism. Especially for which the best antidote to the ills of a highly refined as well as modern civilized world was a return to simple and more primitive life (Cronon, 1995). With a gradual increase in urban population and no frontier, Roosevelt, and Grinnell feared America would lose this edge.

They believed that Americans could cultivate their pioneer skills together with a sense of fair play by doing activities such as sport hunting. Therefore, they can easily maintain the character of the nation (Cutright, 1985; Miller, 1992; Brands, 1997). The Boone and Crockett Club had a great number of influential members, and this was effectively used to great effect in support of these ideals.

Two of the most important and enduring legacies regarding conservation of North America were written by the club members. These were:

1. The Lacey Act (Congressman John Lacey from Iowa, 1900); and
2. The Migratory Bird Treaty Convention (Canadian Charles Gordon Hewitt, 1916).

President Theodore Roosevelt did a lot many things for the conservation of wildlife as compared to any single individual in the history of the US through the institutionalization as well as popularization of conservation and by expanding federal protected lands at large extent (Brinkley, 2009).

Canada did not hold the practices and policies of ownership and management of wildlife as accepted in Great Britain, primarily amongst these being the tie of hunting and wildlife to landownership, and the sale of wildlife as just a commodity in the marketplace.

The fact is more remarkable that some of the negotiators and movers of Canada who were instrumental in the development of novel system of wildlife conservation were immigrants to Canada that is Englishmen.

It seems that at the end of century, when both the countries became cognizant of the wildlife's plight and contended for solutions, then the like-minded elites arose on the sides of the border of both the countries who knew and befriended one another, learned from success and failures of each other and acted on them with understanding and then resolved.

The efforts of Canadians mainly revolved around the Commission on Conservation that was constituted under the Conservation Act of 1909. The commission was chaired by Sir Clifford Stifon until the year 1918. It was comprised of 18 new members and 12 ex-officio members (Geist, 2000).

At the time of the early 20th century, a considerable infrastructure for the conservation of wildlife was in its place. Though, by the phase of the 1920s, it was much clear that the major focus of the system was on the restrictive game laws and that was not much enough in itself to deal with decline in wildlife.

Aldo Leopold, A. Willis Robertson and various other conservationists published an American Game Policy in the year 1930 (Leopold, 1930). It proposed a program of restoration to enhance the legal framework of conservation. They called for a profession of wildlife management with trained biologists, equitable, and stable funding to implement their work as well as university programs in order to train them.

Within 10 years, much of what the policy called for has been realized along with the establishment of the first game management curricula at the University of Michigan and the University of Wisconsin and also, the development of the formation of The Wildlife Society, Cooperative Wildlife Research Units and the passage of the Pittman-Robertson Wildlife Restoration and Duck Stamp Acts.

Such achievements were all founded in the Us initially but a lot many were recognized and mirrored by several Canadian programs and policies.

The following years brought expanded legislation (e.g., U.S. Endangered Species Act and Canadian Species at Risk Act) and programs (e.g., Migratory Bird Joint Ventures and Teaming with Wildlife), but their principles had been set firmly in place. These principles arose amongst the social as well as environmental circumstances that were quite exclusive to the world in their temporal juxtaposition.

4.7. CONCLUSION

Population models are considered to be as the tools that can be very effective to use but are never perfect. The effectiveness of the information acquired from population modeling is related to the data reliability that is generally used to assign definite values to parameters in a model and this highly depends on how accurately the model structure represents the life-history of the study organisms and whether and to what level, stochastic as well as deterministic factors and major management actions that impact the study population are considered.

Models that capture the free-ranging burro or horse life-history, social structure, genetics, and behaviors sufficiently or that encourage the ecosystem processes are probably very complex and need more parameters as compared to simpler models. Though, HMA managers are usually constrained by the absence of that information. As a result, it is quite difficult to suggest some specific frameworks for modeling.

A very suitable framework of modeling or site of models would basically have to encourage the life history; mating system, genetics, social behavior, forage limitation, use of water, forage, and space together with effects of alternate management actions all over the life spans of burro and horse in order to meet the challenges highlighted in the previous paragraphs and also, to include the factors as well as processes that impact the free-ranging equid population dynamics. More probably, different models could be used to address different aspects of the overall problem.

REFERENCES

1. *A Framework for Evaluating the Effectiveness of Conservation Attention at the Species Level*. (2020). [ebook] p. 11. Available at: https://www.researchgate.net/publication/280216748_A_framework_for_evaluating_the_effectiveness_of_conservation_attention_at_the_species_level (accessed on 21 August 2020).
2. Barba, M., Waits, L., Garton, E., Genovesi, P., Randi, E., Mustoni, A., & Groff, C., (2010). *The Power of Genetic Monitoring for Studying Demography, Ecology and Genetics of a Reintroduced Brown Bear Population*. [ebook] Molecular Ecology, p. 14. Available at: https://sciences.ucf.edu/biology/d4lab/wp-content/uploads/sites/139/2016/11/De-Barba-et-al.pdf (accessed on 21 August 2020).
3. *Monitoring of Areas and Species/Populations to Assess Effectiveness of Conservation/Management Actions*. (2020). [ebook] p. 19. Available at: https://www.bioversityinternational.org/fileadmin/user_upload/online_library/publications/pdfs/Crop_wild_relatives/13.Monitoring.pdf (accessed on 21 August 2020).
4. *The North American Model of Wildlife Conservation*. (2012). [ebook] p. 60. Available at: https://wildlife.org/wp-content/uploads/2014/05/North-American-model-of-Wildlife-Conservation.pdf (accessed on 21 August 2020).
5. Ufaw.org.uk. (2020). *Monitoring of Wildlife Populations*. [online] Available at: https://www.ufaw.org.uk/why-ufaws-work-is-important/monitoring-of-wildlife-populations (accessed on 21 August 2020).
6. *Using Science to Improve the BLM Wild Horse and Burro Program: A Way Forward*. (2013). National Academies Press. [online] Available at: https://www.fs.usda.gov/Internet/FSE_DOCUMENTS/stelprd3796106.pdf (accessed on 27 August 2020).

5

Wildlife Conservation: Theoretical and Legal Approach

CONTENTS

It is important to conserve and preserve the environment and the wildlife that is a major part of it. The authorities have realized in the recent past that the wildlife population has been dwindling at a very rapid pace and it needs to be controlled before the time runs out.

There have been various kinds of legal and theoretical approaches that have been proposed and have been adopted by the authorities and the people working in the direction of wildlife conservation, to focus on efficient management of wildlife. This chapter dwells upon those approaches in detail.

5.1. INTRODUCTION

A conscious effort by the human race to protect the plants and animal species and their habitats is referred to as wildlife conservation. In fact, it has two meanings; One as mentioned above, is the protection of both species and its diversity, and second is the wellbeing of animals. It is said that a healthy environment leads to healthy wildlife.

Through conservation education and activities, people's knowledge, attitudes, emotions, and behavior towards wildlife could be influenced. It is an attempt to reconnect humans with the natural world. Wildlife conservation is significant because wildlife plays a critical role in keeping up the ecological balance and contribution to the quality of life for humans.

Figure 5.1. Wildlife conservation is one of the top priorities of the conservationists.

Source: Image by Wikipedia.

Conservation education is defined as the study of man's intelligent use of his natural environment through the development, management, preservation, and renewal of natural resources for his material, cultural, and aesthetic needs to benefit present and future generations. One of the components of conservation education is wildlife conservation.

Since people across the world are concerned with environmental issues, wildlife conservation has also become one of the significant concerns. Saying so, the debate is still on because of the differences in the wildlife conservation concept and the reasonable ideas to conserve the same.

According to some, wildlife conservation means the protection of all animals, and hence they follow the principles of animal. So, which means, such people believe to fight even for live bile bear, also called battery bears, used by some traditional Chinese medicine practitioners, and is also against hunting and related activities.

Moreover, conservation of all animals gives rise to strong public feelings and emotions, which may give birth to excess wildlife conservation activities and as a result of welfare and animal rights. On the other hand, there is a different theory with regard to wildlife conservation.

These people believe that wildlife conservation should be dependent on the scientific approach and method. However, in today's time, people are more inclined towards absolute wildlife protection than scientific wildlife protection.

When it comes to absolute wildlife protection it means the random release of animals, which has given rise to the attack of alien species weakening and preventing the complete process of wildlife conservation itself.

As it can be seen, the attitude of people plays a crucial role in defining, developing, and implementing wildlife conservation management policies. While these policies are being formulated, the policymakers need to address some of the problems that need to be solved while constructing an ecologically-based civilization. These problems are:

- Define wildlife conservation;
- Promote scientific and rational wildlife protection; and
- Strengthen and impartial approach of wildlife conservation by the people.

With the rapid increase in population, the demand for land, sea, and water is expanding too. With this and massive impact on climate change, the conservation and management of wild animals and wild areas is of paramount importance now.

However, people hold different views as to why conservation is important and these reasons can mold conservation strategies in many different ways.

Some of the fundamental elements that direct conservation policy is discussed below with an explanation of how different policies are responsible for ethical confusions and arguments.

This is known to all, that wild animals have always been an important resource for different things like food, fur, leather for human beings. In today's time, wildlife is known to have high economic and cultural value.

Wild animals are seen in circuses, zoos, and wildlife parks, point of attraction in international tourism and are regarded as a source of entertainment these days. Not only that, wild animals are one of the important members of ecosystems on which the human race depends on some of the essential services.

Having said that, sometimes wild animals pose serious threats to humans. For example, animals can be the cause or origin of new human diseases (zoonotic), they can even damage or consume human crops. What is more crucial at this point is, whether the wild animals are resource or threat, how useful are the wild animals to humans. This is what is called instrumental value by environmental ethicists.

Today, while debating on wildlife, apart from instrumental value, other values have also gained momentum. One of them is about the well-being of wild animals (e.g., in terms of animals' flourishing, or suffering).

At the same time, wildlife conservationist are also concerned about preserving wild animals' species or populations, about safeguarding the ecosystem of which wild animals is a crucial part and protecting wild nature as a whole.

In this respect, the wealth of individual animals does not matter because the emphasis in wildlife conservation above is of whole species, ecosystems, and wild nature. It is important to understand the importance to promote the well-being of the ecosystem or the selected species as a whole.

As per the dictionary meaning of 'wildlife,' it means undomesticated animals living in the wild, including those hunted for food, sport or profit. Still, wild animals are not left alone.

Since the last two centuries, a nature and wildlife protection movement has gained momentum in response to protect wild animals and their habitats.

This protection of wildlife comes under wildlife management where, as per the conservation policy, some species are controlled to promote the success of other species.

5.2. CONCEPT OF SCIENTIFIC WILDLIFE CONSERVATION

As mentioned above, different theories and attitudes exist with regard to the conservation of wildlife. And to understand people's opinions and views, it is important to first understand the relationship between humans and wildlife.

Figure 5.2. A wildlife repository.

Source: Image by Flickr.

5.2.1. Relationship Between Humans and Wildlife

Nature and wildness have always been important topics in various eras of world history. During the time of human evolution, wildlife was an important natural resource for humans. Not only that, animal farming has provided all the required materials for the long-term human development. This shows how the relationship between wildlife and humans have always been close yet complex.

5.2.2. Wildlife as an Important Resource in Human Evolution

According to the researchers, Homo habilis i.e., an extinct species of primitive man and Homo sapiens (archaic) i.e., modern human beings as a species increased the meat protein diet by hunting for their food, which resulted in the accelerated development of the brain.

Not only that, meat was regarded as critical for the reproduction and growth of prehistoric humans. Unlike, homo habilis, modern humans i.e., homo sapiens used fired to cook the food which softened the food as well. This resulted in a great change in human structure.

Like for example, since the food was softened, the time taken to chew the food and time taken to digest the food reduced considerably resulting in the change in the jaw and dental structure, such as jaw retraction, smaller teeth, rostrum retraction and smaller crinium visceral.

In addition to the above, man learned about the uses of fire and the consumption of cooked meat protein instead of raw meat which took longer time to get absorbed in the body.

Thus, the discovery of fire and using fire to cook promoted the development of the human brain and body simultaneously. At the same time, hunting also played an important role in improving human cognition and problem-solving skills and also gave birth to the concept of physical fitness.

In nutshell, meat consumption promoted health by reducing disease and at the same time helped to stimulate the brain and other organs.

5.3. IMPORTANCE OF WILDLIFE IN THE LONG-TERM DEVELOPMENT OF HUMAN SOCIETY

It will not be incorrect to say that without wildlife, human life would not exist. Wildlife has played a crucial role in the development of clothing, medical items, experimental models and scientific research. It is a known fact that animal skins and fur are used for clothing and other items for more than 1000 years now and even today they are regarded as symbols of fashion.

Wildlife is also considered as main source of protein hence as a source of nutrition to man. The three revolutions of the human diet that the modern man has seen are: invent of meat and milk production, domestication, and farming of wildlife and the advancement of feed technology.

This helped humans to have more nutritious food with more fat and protein content in the diet which further helped in the development of the human mind and body.

During the onset of homo sapiens, widespread farming and initial development of agricultural produce also increased the quantity and variety in foods, which directly or indirectly contributed to the development of the brain.

However, at the later stages of homo sapiens, primary agriculture, animal farming and the industrial revolution, there has been a change and stability in the structure, habit, and concept of what man eats. Besides the physical form remains more or less the same.

Apart from the above factors that have contributed to the development of humans, medical advancement is also one of the main factors. And again, animals play an important in traditional Chinese medicine both in the form of raw materials and secretions. It also works as a raw material for scientific and medical research.

So far, it is clear that humans and wildlife complement each other, especially when it comes to culture, traditional Chinese medicine, food, hunting, scientific, and medical research and eco-tourism. Hence, it becomes difficult as well as inappropriate to separate wildlife from humans.

It is well proved that wildlife is not only socially but also scientifically important to humans. Hence the need of the hour is to discuss how best wildlife could be protected and the ideology or basis of protection.

5.4. DIFFERENCES BETWEEN WILDLIFE WELFARE AND RIGHTS

Yes, the two words, wildlife welfare and wildlife rights are often used interchangeably, but actually there is a difference between the two which definitely needs clarification.

People who advocate wildlife rights believe in the ideology that there is no difference between humans and animals, which means animals also have desires, feelings, emotions, pain, etc. just like humans. At the same time, animals also possess rights to not being hurt or exploited, so it comes on humans to not use animals in any kind of research or merchant farming.

So much so, according to the advocates of wildlife rights, animal ownership is also considered as wildlife exploitation. Thus, according to wildlife rights, animals should not be treated as food items or used for any kind of entertainment or experimentation.

An American advocate for Animal Rights, Professor Tom Regan, insisted that animals have intrinsic value as they also have feelings, beliefs, preferences, memories, expectations, etc.

For example, in March 2012, China Southern Airlines canceled a shipment of crab-eating macaques (*Macaca fascicularis*) from Guangzhou to Los Angeles because of PETA's (People for the Ethical Treatment of Animals) protest.

On the other hand, proponents of wildlife welfare advocate a responsible use of wildlife to meet human needs. They believe that animals should be treated with the utmost care and are against animal cruelty.

Sound veterinary and animal husbandry, sound research and practices, proper housing, nutrition, disease prevention as well as treatment are some of the factors wildlife welfare advocates lay stress on. Wildlife welfare advocates are not against the use of animals by humans for their own benefit, subject to the condition that it is without unnecessary pain.

Professor Carl Cohen, in the paper 'The Case for the Use of Animals in Biomedical Research" stated that 'animals can't have rights and so animal experimentation can't violate their (nonexistent) rights. Animals can't have rights because they are not part of a community of moral agents, capable of responding to moral claims.'

In today's time, more than 100 countries have established animal welfare regulations, especially in Western countries in Europe and America. In the welfare regulation, animal welfare legislation is been laid down in much detail and explained well.

Some of the acts like mistreatment of animals, inadequate breeding conditions and abandonment of pets will be prosecuted and the offender to face the legal consequences in these countries. In fact, even in legitimate animal experimental organizations, if the animals are poorly or illegally treated, it will result in legal action against such institutions/organizations.

Saying so, it does not state that in above mentioned countries in Europe and America, animals have the same legal protection like human. And, so wildlife rights and wildlife welfare not only differ in definition but also in application and so in reality it is unrealistic and not at all logical and scientific to advocate for wildlife rights blindly.

5.5. DIFFERENCES BETWEEN ABSOLUTE AND SCIENTIFIC WILDLIFE CONSERVATION

5.5.1. Absolute Wildlife Conservation

The attempt to clarify the difference between wildlife welfare and wildlife rights has been made above, however, in reality, sometimes extreme events exist in wildlife conservation. Like for example, the random release of the red-eared turtle (*Trachemys scripta elegans*), which led to a substantial reduction in China's native turtles.

And second is the promotion of going vegetarian to a great extent. As a result, the people who advocate this safeguard and protect animals (any) without any valid reason or even understanding the consequences. For such ultimate decisions, there are two reasons.

Figure 5.3. It is extremely important to conserve the biodiversity.

Source: Image by Wikipedia.

First, namely is the exaggeration and confusion between wildlife welfare and wildlife rights where animals and humans are equals and hence use of animals in any manner is immoral.

Second is when people impose their personal choice on others, such as people who are vegetarians and regard animals as family members strongly oppose the use of animals based on their personal interests. Absolute wildlife conservation being non-scientific and irrational may influence the outlook and approach of the general public concerning wildlife protection.

On the other hand, scientific wildlife conservation refers to the dialectical and unbiased approach to understand the relationship between wildlife protection and its use. Wildlife protection and utilization are indivisible, and if tried to separate, it could actually impact the balance between humans and nature as a whole.

As the name denotes, scientific wildlife conservation calls for scientific use of wild animals and a logical view to protect wildlife. According to this approach, it is not logical to oppose all activities where animals are opposed for the sake of wildlife protection, neither its associate wildlife welfare with wildlife rights because of a love for animals.

Moreover, the advocates of the scientific conservation approach, believe it is incorrect when people try to impose or force their personal choice on the public at large. This results in distortion of the public's approach towards wildlife conservation.

5.5.2. Wildlife Conservation Based on Holism

To understand the scientific wildlife Conservation theory, it is significant to know the relationship between the protection of wildlife and the use of wildlife in a holistic fashion. This means that it has to be understood as a whole and can't be understood separately.

Holistic approach implies that the protection of wildlife and use of wildlife is united and not conflicted. But saying so, people incorrectly believe that wildlife protection and its uses are contradictory.

One of the most common examples is that of animal hunting. The advocates that are against hunting talk about their protection. Their reasons are based on an inadequate in-depth understanding of the subject.

According to the principle of wildlife management, disordered, and erratic hunting does cause severe damage to animals, and on the other hand, systematic, and well-managed hunting can prove to be beneficial to wildlife conservation.

If the hunting activity is well organized, it has quite a few ecological advantages, economic, and social advantages too. As far as ecology is concerned, well-organized hunting can actually help to control wildlife population.

Say, for example, the imbalance ecological takes place when overpopulation of herbivores is more than a low number of large carnivores. The ecological balance could take place through the interference of artificial control of herbivore populations.

Looking at the economic benefits of organized hunting, according to the United States Fish and Wildlife (USFWS) Service, the effect of wildlife hunting on economics was over ＄65 billion in 2002, and direct income from wildlife hunting was $22 billion (Cao et al., 2014).

According to the statistics, hunting offers more employment opportunities to more than 700 people, with one in every 15 Americans engaged in hunting activities.

If it is registered as a company, their tax ranks as 35 in the USA fortune 500 companies. Moreover, the 2011 USFWS report states that 1.37 million people over 16 years participated in hunting activities, with total hunting expenses of ＄33.7 billion.

If one now talks about absolute wildlife conservation, it would refer to the random release of animals. According to this conservation theory, in order to protect the wildlife, organizations, and individuals have planned and provoked people for the random release of animals into the fields. They even buy animals to set them free.

However, this approach has given rise to the illegal trade of animals and has also led to the destruction of the natural environment because of the invasion of foreign species.

And at the end the original intention of doing such thing gets lost. This proves that there is a need for a more structured and professional way to protect the wildlife instead of fervent and inexpert ways. There is a huge difference between the scientific or professional wildlife conservation methods and extreme conservation. As mentioned above, the protection of wildlife and its use are inseparable.

Hence the relationship between wildlife conservation and wildlife exploitation needs to be handled logically. Destruction of wildlife is the opposite of protection which at the end results in ecological imbalance and not the controlled use of certain wildlife products.

5.6. CONSERVATION: LEGAL APPROACH

In today's time, wildlife conservation has become the top priority towards the ecosystem. There are legislations that aim to conserve wildlife. Under the legal framework, general guidelines could be drafted that will help public authorities, individuals, and communities in this regard.

The more specific techniques could be enforced by law, such as species-based conservation, area-based conservation, protection of wildlife from harmful processes etc. Not only this, viable use of wildlife also works towards the conservation of wildlife.

This gives rise to stakeholder's active involvement and contribution towards the reduction in poverty and continuous development. For wildlife management planning, all of the above factors need to be taken under consideration. This will further ensure that communication among species and their habitats are accounted for and there is an opportunity to continuously review and update specific approaches.

Figure 5.4. Trophy hunting has been banned for conserving wildlife.

Source: Image by Wikipedia.

5.6.1. Using a Species-Based Approach

As the name suggests, species approach to protect wildlife species refers to identifying and protecting endangered species. And the good news is that this approach has been accepted as the most appropriate approach to conserving wildlife.

One of the advantages of a species-based approach is that it aims to conserve the species which is not dependent on where it occurs. This helps to define the conservation plan on the broadest scale. According to this approach, it calls for inter-agency and even cross-border initiatives.

The wildlife laws should be focusing on the protection of wildlife and their habitats throughout the world. Moreover, it should just list the protected species because it will defeat the conservation management planning and obstruct the flexibility which is gained through scientific knowledge and changed international responsibilities.

The wildlife law should comprise of a well-defined responsibility, principles, and processes keeping in mind the wildlife conservation.

Legal options

- The authority responsible for wildlife conservation should formulate a well-defined species-based conservation management plan. This plan should include: current status of the species, trade, habitat. Not only that, it should also document all the uses and processes that may get impact the conservation status of the species being discussed.
- In area-specific conservation management planning, it should include definite conservation requirements for the species in question. This means the detailed requirements must be mentioned in the plans like forest management plans, protected area management plans, wildlife reserves, transboundary initiatives, etc.
- It would be essential to list down in the order of higher levels of protection to species whose conservation status is not favorable. Such species are commonly known as endangered species and it comes under endangered species legislation. The same could be separately mentioned and highlighted in wildlife conservation law. This should not be a one-time exercise but a continuous updating would be required basis any scientific information and

as per international listings like for e.g.: CITES (the Convention on International Trade in Endangered Species of Wild Fauna and Flora) and CMS (the Convention on Migratory Species).

- In order to get the intended result, the law should be effective and for that it should have regulatory control to list down all the species and delist, if required. It should also have a detailed regulation on the principle for how conservation of a listed species will take place.

The purpose to list down the species as endangered is to start the exercise for the additional protection requirements to the endangered species which is over and beyond that is applicable to all wildlife in general. The same is applicable to both government and private sectors.

It is important to bear in mind the following to design endangered species legislation.

- Different types of protection plans for the endangered species should be noted down for each level of listing. This is to further make sure that if it is mentioned and listed, further protection must be provided to the species by increased disincentives to poaching and incidental take.
- The conservation aims to be accomplished before the delisting of any species must be clearly mentioned.
- Wherever it would be authorized, the treatment of the independent population of the same species should be allowed in terms of status and trends.
- The responsible agencies/forces are required to cultivate a detailed species recovery plan. For this, they should gather all the information and consult the national scientific authority, local government agencies, and the public in general.
- It is important to ensure that the legal parameters for listing and delisting are based on the scientific approach. This approach is against and totally opposite to the non-scientific approach,
- The definitions and terms used in the legislation should be simple and clear focusing both on status and trend in the population of the species (rare vs. threatened);
- It is imperative to establish a transparent process to list and delist. For the same, scientific research information, any public notice received in this regard and comment process will be required.

5.6.2. Using an Area-Based Approach

In this approach, the principle takes its lead from international conventions. It focuses on an important element of wildlife conservation which includes habitat protection. However, most of the time the focus on habitat generally ignores political boundaries and thus forces a degree of local, regional, and international cooperation.

Area-based approach to managing wildlife find expression within domestic legislation in three ways:

- To create a protected area system by order which comes under protected area legislation, this would include all the area that is recognized as a critical wildlife habitat;
- To order the primary wildlife management authority to specify areas outside the protected area system. This act should benefit from wildlife specific protection measures; and finally;
- To order to establish an area-based hunting scheme. This will enable only specific areas open for hunting and will also set a benchmark for their designation and management; and
- It is important that the law remains flexible in order to continuously update the list of protected habitats depending upon any new scientific knowledge, local requirements or any change in the international obligations.

5.6.3. Involving Local Stakeholders in Wildlife Conservation

It has always been acknowledged that local communities play a vital role when it comes to manage the local resources. It can be easily done by empowering stakeholders and making them accountable for their actions.

If there is no support from the local communities, then all the hard work to protect the wildlife put in by the short-staffed wildlife management officials who are poorly paid as well, goes in vain. In the absence of any support from the local communities, it lessens the interest of the local communities to abide by the law and also prevents them from insisting on the compliance of outsiders, even if they are government officials.

Hence it is important to understand the importance of local communities who actually literally live with and are also impacted by the use and conservation of biological diversity, not to forget their contribution towards the conservation.

Legal options

- To manage natural resources, local stakeholders which includes native and local people should be involved. For their involvement and to keep their interest alive, fair compensation should be provided for their hard work both in the form of monetary and non- monetary ways.
- Not only that, but some additional economic motivation should also be promoted for the local stakeholders which include native and local people who are actively taking interest in the management of biodiversity components.

These economic benefits can be in the form of job opportunities for native people, training, and capacity building, fair distribution of profits amongst the local stakeholders and outside stakeholders.

- Where the foreign investment is greatly involved, the fair share of the benefits should be provided to the local people.

Local people should be involved in the decision-making process and wildlife management of protected areas (PAs). If the traditional use of the local people to utilize the resources is in sync with the conservation objectives, the same should be allowed in the protected area. At the same time, if there is a constraint to traditional use for wildlife conservation purposes, then the same should be offset.

- Fine to be allocated in case of poaching to local community members that contribute to detect and stop illegal activities.

5.6.4. Protecting Wildlife from Harmful Processes and Negative Impacts of Other Land Uses

Wildlife conservation is much more than just protecting the species of their habitats from such activities that directly affect the wildlife like off-take and trade of wildlife. It also refers to safeguarding the wildlife from all such activities that may indirectly hamper the animals in a negative way.

Such activities that indirectly impact wildlife involve industrial development, construction everywhere, increase in tourism, mining operations. All these activities disturb the wildlife species directly or indirectly and also destroys their habitat.

Not only this, the uses of land for forestry or agriculture also impacts wildlife. Different pieces of legislation regulate in different and maybe conflicting ways impacts wildlife. This is due to the fact that the actual effect of human actions or even potential effects may disturb the ecosystem.

According to the Convention on Biological Diversity, countries are supposed to find out and take necessary actions to control all possible sources of detrimental impacts on biodiversity. It should also conduct environmental impact assessments of projects that are supposed to have critical adverse effects on biodiversity.

Therefore, it is significant for the wildlife laws to take all of the above into consideration when creating a law for the conservation of wildlife. It should also be providing tools to detect and reduction of the impact.

5.6.4.1. Legal options

Law can play a crucial role at several stages in safeguarding and protecting wildlife from injurious processes:

- The evaluation of any such processes that may be injurious or hurtful to animals which can be done usually through an environmental impact assessment. It should specify all the steps and minimum requirements for example the need of considering all alternatives-whether administrative or other activities which is directly or indirectly effecting wildlife and their habitats.
- The public should be allowed to appeal for the assessment as per law. And if required by public authorities, people should be allowed to contribute in the form of any information or comments in the said assessment.
- The legal consequences of such assessment should also be specified in the law, such as, whether the expected negative effect will hamper the conduct of the proposed activity altogether, or whether the activity would be carried out but only as per the specific requirements necessary to minimize the negative effects or remedy to them.
- Another measure or step that could be taken is to levy restrictions on all such activities that prohibit all those activities that may cause permanent loss to the environment.

If general environment legislation has already specified the rules and regulations that are applicable to wildlife, it is still recommended for the wildlife law to explain the link with general rules on environmental effect assessment, to avoid any legislative conflicts and problems in understanding.

- Law should consider all the likely negative effect on wildlife of competing land uses (such as restrictions and other requirements under legislation regulating the forestry, agriculture, mining, and tourism)
- Constituting a general responsibility for the reduction of harmful activities;
- Keeping a record of essential threatening processes and requesting a recovery plan for affected wildlife.

5.7. INNOVATION ADOPTION THEORY

Innovation Adoption Theory is a theory by Rogers that categorizes people who adapt innovation into various successive categories. It is actually based on the idea that there are certain people who are surely more open for adaptation than rest.

This theory depicts how innovations are advanced, adopted, and incorporated into everyday organization's process or by individuals within the organization and is encircled by an extensive body of well-developed literature. (Rogers, 1962; Moch and Morse, 1977; Tornatzky and Klein, 1982; Leonard-Barton and Deschamps, 1988; Davis et al., 1989; Cooper and Zmud, 1990; Prescott and Conger, 1995; Taylor and Todd, 1995; Karahanna et al., 1999; Gallivan, 2001).

This theory has given birth to many other theories and foundations to explain the adoption of innovation into any organization and the distribution of innovation across them in different ways and from different angles (Daft, 1978; Cooper and Zmud, 1990; Gallivan, 2001; Courchamp et al., 2006).

However, the focus of this theory is completely on the information technology organizations and innovations. It is yet to make its way into other area. Nevertheless, the research and foundations which is developed for IT organizations bring forth many favorable circumstances to study innovation in wildlife management and wildlife conservation.

Gallvan's theoretical framework associate's theory on individual innovation adoption with theory on organizational innovation implementation. This thus create a composite framework that more strongly takes into account the complexity of interaction between managers, appointed strategies, and other inherent factors of organizations.

In the framework, the innovation adoption process is analyzed and cut up further into three main parts: the main authority decision, secondary adoption and adjustments and the organizational consequences. The main authority section includes the decision taken by the top-level i.e., Management level to accept an innovation or not.

The same is simple and well understood. Secondary adoption and adjustments comprise of major complications of the framework and is further broken down into three constructs: managerial intervention, subjective norms and facilitating conditions.

Together these three constructs dictate the occurrence of the secondary adoption process and adjustment stage of innovation adoption and is the focus of this study too. Third and last is the organizational consequences, which actually explains the result of innovation adoption on the organization adopting it.

Managerial interference denotes that the decisions are made by the managers as to whether they are going to adopt the innovation or not and also, they take actions to make sure the adoption process is a success (e.g., training, resources, support, etc.).

Subjective norms refer to individual's judgment about the innovation, which also includes the role within their responsibilities and whether innovation should be accepted or left alone. Gallivanstates, "facilitating conditions 'is a broad category that captures other factors that can make implementation more or less likely to occur.'

These can very well include the features of the innovation itself, the organization or of individuals and also keeping in mind the environment, in which a particular innovation would operate.

The arrival of advanced technology into conservation and environmental sciences brings forward many new opportunities. However, challenges do not leave them either. A lot many technologies are still new to conservation problems, plan to use advanced technology for a given problem are prone to poor planning and techno-fixes which do not work as intended and end up wasting time and resources as well.

As presented in Gallivan's framework if the principles of innovations and adoption are followed diligently to conserve and manage wildlife, it presents one potential path to critically assess a technology's applicability to a conservation problem and its probability of successfully assimilating into the system.

Already, conservation is extremely resource limited and as more and more opportunities arises for technology in conservation it becomes apparent, practitioners will require tools to assess a potential solution to a given scenario before executing full resources to the development and implementation of that solution. Innovation adoption theory adapted from IT management to conservation and wildlife management may provide such a tool.

5.8. LIMITATIONS IN WILDLIFE LAWS

No doubt Law is one of the many tools which is used to accomplish the objectives of wildlife conservation. The formal organization of law operates within a broader context which includes informal organizations. Much smaller space is taken by international law within the conservation framework.

The reason for the same is public international law, under which sovereign states achieve agreements on a voluntary basis in an international legal order that is deficient of centralized legislative, executive, and judicial powers which are applicable to domestic legal orders.

Under such circumstances, broadly advertised purpose to end and reverse the loss of biodiversity already looks difficult to meet in reality. Agreement is generally not perfect and failure of application and administration hampers even the refined legal instruments (Bowman et al., 2010; López-Bao et al., 2015; Wandesforde-Smith, 2016; Chapron et al., 2017).

The difference in ideologies between the parties of the agreements is yet another problem. The same is clearly seen in the frequent debates within CITES over the proportionate weight to be given to rigid protection policies versus continuous use of elephant and rhinoceroses (Couzens, 2014; Wandesforde-Smith, 2016).

The same kind of struggle is visible within the International Convention for the Regulation of Whaling (ICRW) warns this regime flawed (Couzens, 2014).

To be compelling and persuasive, international legal instruments must mention clear and sufficient commitments, draw the attention of acceptable parties and ensure a great deal of compliance (Bowman, 2000).

These international legal instruments differ in the intensity to which these instruments meet these criteria. Wildlife treaty negotiations are always affected by the possible tension between the parties and the other two criteria, and many treaty texts reflect later compromises.

For example, though almost all the states across the globe endorsed the Convention on Biological Diversity i.e., CBD, however, in the legal aspect, most of its responsibilities, is limited to "as far as possible and as appropriate."

Also, there is an absence of an efficient compliance tool in CBD. There are many agreements that provide country-specific exceptions where it grants reservation. This means that a state when it becomes a party or when any new responsibility is endorsed, limits the scope of the agreement versus that state.

For example-the restriction to the zero quota for commercial whaling lodged by Iceland upon rejoining ICRW in 2002 and the reservations regarding wolves (*Canis lupus*) filed by 14 Bern Convention parties.

Basically, international law cannot achieve more than what the world's distinct and fluctuating national administrations and eventually the community they serve, requires or what these communities are able to implement it practically.

As it is, the international wildlife law has not been able to satisfactorily provide the solution to biodiversity loss. Moreover, the hesitation of administration and communities to enforce long term and enforceable restraints on economic development and other human desires.

It is important to understand what international wildlife can administer in reality, mainly because some of the goals may be supported by legal enforcement such as trade agreements. However, wildlife conservation success is not only dependent upon international wildlife law only but is also affected by international law decisions on subjects like climate change, crime, economic integration, fisheries, pollution, and trade.

5.9. OPPORTUNITIES IN WILDLIFE LAW

Despite of everything, the need of the hour for an effective conservation of species and ecosystem as a whole is involving people and communities from different countries and their long-term dedication and devotion.

Though there are lot of limitations associated with international law, but still it is the most important tool to make this happen. As far as the temporal aspect is concerned, international agreements progresses at a very slow pace and once states are part of these agreements, then they remain the part of the same.

And hence, such agreements can propose a legal buffer against the election-cycle swings of national governance. So that means, generally speaking, international law acts as a moral guide and keep emphasizing on the obligation towards conservation made by governments and the public in general.

In the past, wildlife conservation solutions presented by International legal instruments are:

1. Marking or labeling of protected area in accordance and agreeable to international constraint;
2. National legislation has been instigated to regulate the exploitation of wildlife;
3. Wildlife conservation has been given priority on governments list of things to do;
4. Technical guidance as adopted by COP's and other treaty bodies have been integrated into national action plans and legislation;
5. Data has been very well organized;
6. Governmental and Non- Governmental stakeholders' interest is seen in their increased cooperation;
7. Conservation actions have received direct help through treaties funding system; and
8. Where any injurious developments have been taking place were obstructed or conservation action against the harmful development was taken when governments were challenged with their international commitment in national or international court proceedings or through compliance tools.

For example, in spite of few restraints and not so perfect compliance record, the Convention on International Trade in Endangered Species of Wild Fauna and Flora (CITES) apparently contributes to the conservation of species injured by trade.

In 1975, in the act of conservation of jaguars and other South American felids, their condition was improved after CITES boycott on trade in their pelts.

Another instance is the European Unions' (EU's) 'Nature Directives' are comparatively effective conservation instruments resulting from both the clear limits they impose on EU member states' discretion and the special nature of EU law, entailing powerful enforcement options at national and European levels.

Not to forget, Natura (2000) protected area network where many species were benefitted when their habitats were protected and from limitations placed on their exploitation. Similarly, the recent international agreement in regard to the Ross Sea ecosystem where it is agreed to assign the great area of it as a marine protected area.

Even administration with low budget, such as the Agreement on the Conservation of African-Eurasian Migratory Water birds (AEWA) and the Bern Convention on the Conservation of European Wildlife and Natural Habitats, can contribute positively and sometimes show that they have the power to make things work.

Public in general agrees to Bowman and colleagues (2010) when they stated, "cause for optimism in the extent to which international wildlife law is permeating national policy discourse, legal instruments, and (slowly) judicial decision-making."

International wildlife law is increasingly invoked before national courts-generally supported by tools like the Aarhus Convention on Access to Information, Public Participation in Decision-Making and Access to Justice in Environmental Matters.

Moreover, domestic legislative application and administration are very closely examined by international agencies. They keep observing and advocating compliance with international obligations.

Moreover, the increasing power of international courts, as was depicted by the 2014–15 Serengeti highway rulings by the East African Court of Justice; the crucial role of the Court of Justice of the EU in enforcing the Nature Directives; and the increasing number of cases involving wildlife law brought before the International Court of Justice are all visible.

5.10. MAKING THE MOST OF INTERNATIONAL WILDLIFE LAW

The extent of the opportunities presented by international wildlife law is determined by the number and different types of legal instruments that exist for the same. The benefits of international wildlife law can be maximized in various different ways.

There are different officials playing different roles like civil servants managing treaties, scientists taking care of monitoring and involved in advisory bodies and different NGOs playing critical roles in intergovernmental agendas and persuading or compelling governments into compliance.

In general, a strategic approach is the best approach. Just spending in the implementation and improvement of an international establishment may not be the optimum approach to spend the limited conservation resources.

Also, there is no one type of approach but a mix of approaches and tools in international wildlife law instruments. To optimize the impact of conservation, different approach in different conditions is the need of the hour.

And if there is a support instead of oppose, better results can be achieved. Because defiance generally weakens the long-term support for the legal framework itself.

However, in few cases, sometimes any dispute or opposing approach becomes important to ensure compliance even though international agreements are expected to annul national, regional, and local interests when there is a conflict with the agreement's objectives.

Legal experts play a specific role in wildlife law where they are supposed to combine conservation professionals and expertise from other disciplines too. Situations arise when socioecological professionals work to improve the utilization of conservation law and where legal information works to improve biodiversity conservation.

This becomes a way sometimes to make the people who are taking care of conservation sound that they can use legal tools. The books by Bowman and colleagues (2010) and Gillespie (2011) provide good starting points.

There is another approach where national authorities and committees can be reminded to plan the domestic conservation laws, policies or any strategy of the often-little-known wealth of detailed guided adopted by COPs.

And on the other hand, sometimes more detailed and customized legal exercises will be required. From the past experiences, it is very clear that even seemingly ambiguous treaty provisions can have surprisingly far-reaching results when understood in the context of objectives of treaty, interpretive advice adopted by the parties, and/or scientific knowledge regarding a particular conservation issue.

Legal procedure, however, still somewhat remains unfamiliar feature within the multidisciplinary conservation literature and conservation practitioners, in spite of its extensive acknowledgement to wildlife conservation.

International law research methodology comprises mainly of the recognition, analysis, and implementation of legal instruments, including their interpretation as per the prescribed format in the Vienna Convention on the Law of Treaties.

Such reasoning and scrutiny qualify in utility when combined with understanding regarding the ecological, socioeconomic, and cultural elements involved.

It is actually worth a talk when conservationists from other disciplines join hands with international wildlife lawyers to explain and make clear the significance of international tools for given issues: to determine any gaps and difference in the legal framework.

And most importantly, to establish and seek a path towards improvising the application of the law and where required, the law as a whole. This includes working on people's understanding of the role of law regarding conservation disagreements and further understanding as to the time when it is best to go legal rather than taking a more cooperative approach.

International wildlife law litigation is known to be multi-disciplinary attempt. Other similar attempts are wildlife regimes' technical and advisory bodies, International Union for Conservation of Nature (IUCN) Specialist Groups, and Society for Conservation Biology (SCB) policy committees.

For example, the plan to develop AEWA's Implementation Review Process (established by AEWA Resolution 4.6) originated from a lawyer, but the agreement's multidisciplinary Technical Committee is involved in assessing whether proposed cases are appropriate, and initiation of the process may result in onsite assessments by multidisciplinary teams of experts.

There is another instance where the Scientific and Technical Review Panel of the Ramsar Convention has a specific multidisciplinary membership and is currently presided by an international wildlife lawyer.

5.11. CONCLUSION

Wildlife has an important role to play in human growth. The wildlife is an integral part of the ecology, which provides numerous resources for various kinds of commodities and services. It contributes to the pharmaceutical industry, the textile industry, the food industry, to tourism and other such domains as well. The humans need to understand that there is a symbiotic relationship that exists between the wildlife and the humans.

The legal approach involves that the approach is based on the species on which it is focusing, the area in which the species needs to be conserved, the locals involved in the conservatory measures and on the prevention of the wildlife from the harmful processes.

REFERENCES

1. (n.d.). *Innovation Adoption Theory in Wildlife Management and Conservation: Initial Insights from a Case Study of Machine-Vision and Cape Buffalo Management in South Africa*. [ebook] Available at: https://wildlifecollege.org.za/wp-content/uploads/2017/11/E.Walker_OxfordDissertation_FINAL_PDF.pdf (accessed on 21 August 2020).
2. Gamborg, C., Palmer, C., & Sandoe, P., (2012). *Ethics of Wildlife Management and Conservation: What Should We Try to Protect? | Learn Science at Scitable*. [online] Nature.com. Available at: https://www.nature.com/scitable/knowledge/library/ethics-of-wildlife-management-and-conservation-what-80060473/ (accessed on 21 August 2020).
3. Morgera, E., & Wingard, J., (2008). *Principles ForDeveloping Sustainable Wildlife Management Laws*. [ebook] Available at: http://www.fao.org/3/a-bb108e.pdf (accessed on 21 August 2020).
4. Trouwborst, A., Blackmore, A., Boitani, L., Bowman, M., Caddell, R., Chapron, G., Cliquet, A., et al., (2017). International wildlife law: Understanding and enhancing its role in conservation. *BioScience*, [online] *67*(9), 784–790. Available at: https://doi.org/10.1093/biosci/bix086 (accessed on 21 August 2020).
5. Zhou, X., Wan, X., Jin, Y., & Zhang, W., (2016). Concept of scientific wildlife conservation and its dissemination. *Zoological Research*, [online] *37*(5). Available at: https://www.ncbi.nlm.nih.gov/pmc/articles/PMC5071339/ (accessed on 21 August 2020).

6

Practical Management of Wildlife

CONTENTS

In the chapter practical management of wildlife, the overall management that needs to be done for preserving wildlife is discussed. The chapter explains the various major communities' approaches for wildlife conservation.

Some of the key approaches are Integrated Conservation and Development Projects (ICDPs), Community-Based Conservation (CBC), and Community Based Natural Resource Management (CBNRM). In addition, the chapter also highlighted management practices for enhancing wildlife habitat.

The chapter also discusses the sustainable wildlife management. Some key methods adopted in the sustainable wildlife management are vaccination, physical separation, direct culling of wildlife population, and changing behavioral patterns of farmers or hunters.

In the end, the chapter explains the practical aspects of enforcement which includes enforcement costs, enforcement budgets, and the role of fines, and the multiple layers of enforcement, fairness, and conflict.

6.1. INTRODUCTION

It is generally seen that wild animals have always been a critical resource for human beings. Historically, food, leather, and fur, were essential resources for the survival of human beings - more recently, wildlife has assumed high cultural and economic significance.

Wild animals are high source of entertainment in circuses, wildlife parks, and zoos, they form a central attraction in promoting international tourism, and they are crucial members of ecosystems on which humans rely on their essential services.

In a similar way, wild animals can be seen as a major threat to the life of human beings; for instance, they can be sources of new human diseases (zoonotic), and they can consume or damage crops which are originally produced for the consumption of human beings. What matters here, whether as resource or threat, is how useful—or otherwise—wildlife is to human beings. Environmental ethicists often call this instrumental value.

In modern debates about wildlife, however, there are other values which have become increasingly important. One major area of focus is on animal welfare-the overall wellbeing of individual wild animals (e.g., in terms of animals' suffering, or flourishing).

There are various other concerns about protecting populations or species of wild animals, about guarding the ecosystems of which wild animals form a part, and about shielding wild nature itself (Sandøe and Christiansen 2008).

The overall wellbeing of individual animals is not of high importance where ecosystems, species, or wild nature is emphasized—indeed, painful predation may be comprehended as encouraging ecosystem health, or as employing the right kind of selective pressure on a species as a whole.

Although the basic notion behind the concept of "wildlife" is usually taken to mean animals not bred or governed by humans, increasingly, wild animals are not ignored altogether to live their own lives (Gamborg et al., 2010).

In response to continuous increase pressures on wild animals and their habitats, a nature and wildlife protection movement has grown over the past few decades. Often this protection has over a period of time transform into active wildlife management, where some species have been controlled as part of a policy to endorse the victory of other species.

Human attitudes towards wildlife and wild nature have been since past ambivalent. With reference to the cave paintings, it can be concluded that prehistoric societies of hunters and gatherers seem to have perceived wild animals not only as a source of food and fur but also as objects of reverence.

And while a leading strand of the Judaeo-Christian tradition recognizes animals purely as human resources, other Christian traditions—such as St Francis' celebration of animals as " sisters and brothers"—infer the importance of animals in a completely different way (White, 1967).

It is important to note that the ideas about wilderness have equally been ambivalent and complex: wilderness has both been implied as dark, fearsome, and chaotic, but also as unsullied, a place of spiritual sanitization and sublime attraction.

In the past few years, many Wildlife Management Practices (WMPs) have been initiated with the objective to manage wildlife and their habitat. In this section, the potential impact of WMPs on wildlife habitat and populations have been discussed.

The WMPs are grouped with reference to their category of practice (Population management, Habitat management, Stream/ Pond Management, Additional management practices aimed at Urban areas) and listed in alphabetical order within each grouping.

Contestants should be acquainted with the WMPs and able to identify which WMPs might be advocated to enhance habitat or adjust populations in the ecoregion used for the Invitational (or local or state contest). There are various practices that have been used commonly in various ecoregions, but not in others. It is important to gain enough knowledge as one can about any WMP before suggesting it to anyone. Some.

WMPs may seem incongruous. Landowner aims, as well as explicit info given by contest organizers, must be deliberated to decide the adequacy of WMPs. Some WMPs are not pertinent in all ecoregions, even though certain of the species may be the same. Current conditions should be considered when deciding if a WMPs needs to be applied within the next year.

However, the various key advantages of a WMP may not be realized for years. For example, planting trees in a field to provide a favorable environment for eastern gray squirrels or acorns for wood ducks is a good practice, but their advantage will not be able to realize over certain years. In this manual, budgets, and costs are not taken into consideration while recommending practices. However, in real-time situations, wildlife managers must take into notice economics when planning and recommending WMPs.

6.2. ORIGINAL IDEAS OF MAJOR COMMUNITY APPROACHES

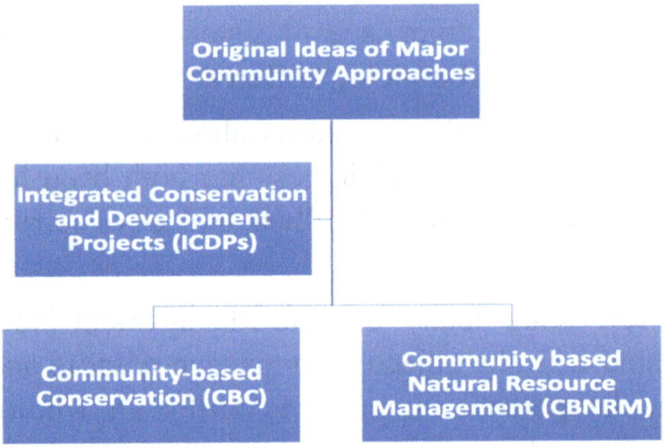

Figure 6.1. *Original ideas of major community approaches.*

6.2.1. Integrated Conservation and Development Projects (ICDPs)

The conceptualization of Integrated Conservation and Development Projects ICDPs is done in Wells et al. (1992).

ICDPs are considered to be "new tactics and strategies aimed at protecting area management that further is concerned with addressing the needs of nearby communities by stressing local individual's engagement and by combining conservation with development."

The key objectives are to protect the public areas, and the goal is biodiversity conservation inside of them. The concept of ICDPs is one of the initial efforts aimed at forming at a new conservation strategy, and there are two originalities at that time.

It is usually seen that the extraction and deprivation of natural resources within the protected areas (PAs) are carried by development projects, prohibited hunting, extension of farmlands, and logging so on. In a way to address these kinds of situations and to accomplish the key objective of ICDPs, "local-participation" is demanded; this is the preliminary essential feature.

This concept has been considered as a process which helps in empowering local people in a way to make sure that they can control resources as well as their lives. The second unique characteristic is the integration of rural development and biodiversity conservation. Before ICDPs, it is largely believed that said, conservation, and development both were separately pursued. ICDPs are basically aimed at combining these two things.

As components of ICDPs, buffer zones, shielded area management, and local economic and social development are mentioned, and then, it is said that the latter one seriously differentiates ICDPs from the preceding projects. Although the development components play very critical role for ICDPs, it has been designed primarily to accomplish the goal of biodiversity conservation.

It is largely believed that the practical methods of "local participation" is regulated or governed by this strategy. Compensation and substitute are described as chief measures to bring economical advantage to the communities, but then, the use of local resource is seen as something that need to be prohibited or regulated for sake of biodiversity conservation. In addition, the people are considered as passive targets of outsiders' conservation education.

6.2.2. Community-Based Conservation (CBC)

Figure 6.2. Endangered animal of a particular area.

Source: Image by Wikipedia.

The concept of community-based conservation (CBC) was formulated by the Western et al. (1994). The key objective of the CBC is that "local engagement and participation in decisions making and benefits could result in minimizing the hostility toward conservation efforts" (Western and Wright, 1994).

The basic fundamental of CBC is to create local conservation initiatives that comply with the interests of outsiders (Western, 1994a).

The achievement of CBC is monitored by the real progress of conservation. In other words, "achievement of CBC eventually should be assessed by how deeply the effort is integrated in each community's goals and how efficiently its supporters' efforts sustain it."

It is worth noticing that CBC has a wide range of measures (ibid: 507), but the critical thing is neither the nature of methods nor the initiative. One most important thing in this is that whether it produces tangible local benefits (Western and Wright, 1994). It is well known that CBC takes into consideration local benefit because it is of the opinion that local people are those "who bear the cost of conservation."

The carelessness to that cost is thought to be the critical thought of why people have often criticized to conservation initiative. One of the major reasons why CBC is not excessively worried about the initiative and strategies is that it assumes today's conservation cannot be possible only by the local communities.

Of course, there is a need on part of government as well as other organizations to abandon their traditional top-down approach, but the discussion includes suggestions or methods for such organizations to play active, supportive, and facilitative roles.

As far as communities are concerned, CBC cautions of having excessive expectations. It believed that the privileges of the local people should be discussed with capabilities and responsibilities.

When it is said "local skills and initiatives undertaken by the community members must become the driving force of conservation" (Western, 1994), it is clear that CBC thinks the local people possess the ability and in fact, should take part in conservation activities with their abilities.

Concerning local participation, CBC emphasizes the role that education plays toward the local people, it has more positive evaluation of the local praxis.

For CBC, which tries to form "aspiration," it is important to note that the people are not passive objects but rather subjects who can play the critical role in local conservation activities, even if they require some kind of assistance from outsiders.

6.2.3. Community-Based Natural Resource Management (CB-NRM)

Figure 6.3. Conservation of a particular area.

Source: Image by Wikimedia commons.

Around the globe, the term, CBNRM has been accustomed to multifarious resources. However, According to Turner (2004), in the field of wildlife conservation in South Africa, CBNRM has been used with a characteristic meaning.

To check the current conceptualization of CBNRM for wildlife, there's a reference to a book (Child, 2004a), written by the chairman of Southern Africa Sustainable Use Specialist Group (SASUSG) in this section.

The essential beliefs and experiences of CBNRM are that "conservation will be most effective when placed in the hand of landholders, made responsible by sensible institutions that empower landholders and align and internalize costs and benefits at the level of land units" (Child, 2004b).

In Zimbabwe and Zambia, the wildlife practice is primarily adverted for the conceptualization and modification of CBNRM because of their "outstanding economic and ecological success" (Jones and Murphree, 2004).

Sustainable use as a conservative paradigm, economic instrumentalism, devolution, and collective proprietorship which are the common features of Zimbabwe and Zambia are adopted by CBRNM. Along with this, CBNRM has "the price-proprietorship-subsidiarity paradigm."

This signifies "if the resource is valuable (price), if this value is captured by landholders (proprietorship) and if the principle is followed that no management action, decision or benefit is arrogated to a higher level when it is better and more appropriately conducted lower level (subsidiarity)." A greater emphasis is put on the market economy than ICDPs, CBC, and even CC by the CBRNM, based on this archetype.

In the elucidation of "economic instrumentalism," it is acknowledged, "where wildlife cannot be made economically competitive, its displacement by other forms of land use must be accepted" (ibid: 65).

CBNRM intends "the switch from fortress to incentive-led conservation" (Child, 2004c), and the emphasis on the local benefit is akin to ICDPs and CBC. Whereas, the incongruence is that CBRNM measures benefit in respect of the market economy.

6.3. MANAGEMENT PRACTICES FOR ENHANCING WILDLIFE HABITAT

Figure 6.4. Saving our native prairies.

Source: image by U.S. Fish and Wildlife Services.

Described below are the most prevalent habitat management practices for wildlife. The descriptions are succinct and general. For more information regarding the appropriateness of practices best for your property, ideally, contact a specialist in wildlife management who will assess the situation.

6.3.1. Brush Piles

In this management practice, Resting/escape cover and den sites for wildlife are provided by assembled piles of brush, known as brush piles. Brush piles are used for various purposes.

For example, Eastern cottontails and other small mammals use them for cover, they are used by Songbirds for perch sites, especially if the piles are located near feeding or nest sites. Additionally, they may be used for feeding and resting by amphibians and reptiles, given that the brush piles are adjacent to a water source.

The best brush piles for wildlife start with the largest materials like pole-sized logs at the bottom and culminate with the smallest materials like small limbs and shrubs at the top of the pile. The arrangement of materials is such that the brush pile is slenderly lifted above the ground.

As a consequence of which, animals find it easier to get under the brush pile and into cover. The rate of decay of the brush piles is also decelerated by placing the largest materials at the bottom of the pile.

In the vicinity of food sources, brush pipes are placed, and also in places where low cover for wildlife may be scrimpy or non-existent. Good areas to build brush piles are forest openings, forest edges, and timbered areas because of the variety of types of wildlife feed in openings and along forest edges and due to the presence of piles this wildlife can get extra protection.

6.3.2. Controlling Noxious Weeds and Non-Native Invasive Plants

Invasive or aggressive plant species are generally easily established. It is easy to grow Invasive or aggressive, as once they've grown, they then expand beyond those areas of their initial boundaries. Non-native species that can out-compete native species and dwindle the diversity of natural plant communities are known as invasive species. Wildlife, livestock, and humans may disperse them. Humans intentionally or inadvertently introduced many of them.

Multiflora rose, Japanese honeysuckle and purple loosestrife are some examples of invasive species that may surpass native plants in Pennsylvania. Even though some of these species are advantageous to wildlife, they can create problems and have limited value for most wildlife in the long run.

Because of this, there are various experimenting with a variety of techniques that can be employed for controlling the invasive species and then replacing them with native plants.

Lots of benefits have been obtained from a variety of native plants and plant communities which have evolved from wildlife species in areas such as Pennsylvania. Which we can then say that by administering the growth of non-native plans, biological diversity can flourish.

In Pennsylvania, some species of invasive plants such as kudzu vine, mile-a-minute vine, multiflora rose have been categorized as noxious weeds and it is prohibited to plant them. It is highly recommended to contact local resource professionals to find out as much as possible information about which species may be problematic in case you are not sure which particular species may be invasive.

Natural Resource Conservation Service personnel, Penn State Extension agents, Bureau of Forestry service foresters, and private natural resource professionals can provide insight into the control of non-native species.

6.3.3. Creating Snags

Snags are referred to as dead or partially dead standing trees that assist in a variety of wildlife. Snags provide cavities for perches for hunting and displaying, nesting, and resting, and a sufficient supply of food for insect eaters.

In Pennsylvania, it is reported to have at least 20 species of mammals and over 35 species of birds that use snags at some point in their life cycles. Moreover, many species of amphibians and reptiles also use the cavities in snags.

It is quite obvious that different species of wildlife prefer different sizes and types of snags in a variety of habitats. According to a report, some species prefer hard snags (dead or partially dead trees with fairly sound wood and some limbs remaining) while other species prefer soft snags (also known "punky," in advanced stages of decay, and rarely with limbs).

Barred owls and wood ducks are one of those species that require large snags due to the fact they need large cavities in which to nest.

On the other hand, tufted titmouse once included as a subspecies but is now considered a separate species, will forage and nest in cavities inside smaller snags. To provide a variety of species, many landowners try to manage several types and sizes of snags.

Retaining existing snags is perhaps the best method to provide snags for wildlife so that they will not create a critical situation for people using the nearby area for outdoor activities like cutting firewood or hiking. There are several guidelines suggested for the sizes, types, and numbers of snags that are most suitable for wildlife.

A reference is available free of charge from your county extension office where details about snags can be found in *Dead Wood for Wildlife* (number 7 in the Pennsylvania Woodlands series). When the distribution of snags or abundance is inadequate or if particular types of snags are desired, snags can also be "created."

Creating snags implicates deadening trees so that they remain standing. Success depends on the method used which are the current health of the individual tree, the tree species you are trying to deaden, and the specific site features such as the presence of forest pests that may stimulate the tree's death.

Retaining or creating snags is often included in other habitat management practices. For example, if clearing is intended to create an opening, some of the trees that could be removed while clearing could instead be deadened and left standing for use by wildlife.

In another instance, if a forest-edge cutting or a tree and shrub release is planned, some of the trees that would be removed can instead be left standing and deadened.

6.3.4. Establishing Permanent Vegetation for Wildlife

On some traits, shrubs, herbaceous plants and trees have been planted to provide privileges for wildlife. In most of the cases, the plants selected provide food or cover or even both.

6.3.5. Fencing and Tree Shelters

During the process of establishing new plants for wildlife, it is sometimes essential to preserve plants from browsing and other damage until the plants become well established. In Pennsylvania, some young plants include cottontail rabbits, white-tailed deer, and meadow voles may have been damaged by wildlife species.

Cottontails girdle and meadow voles (chew the bark and cambium layer off the stem near the bottom of the plant) woody-stemmed plants while deer browse stump sprouts seedlings, and shrubs. In rural areas, livestock may browse or crush young plants.

As a result, fencing or seedling protectors may be required to protect naturally regenerating or newly planted trees and shrubs. Fencing is also used to eliminate livestock from wetlands and streams (see Streambank fencing).

A variety of fence types are available such as high-tensile strength, woven wire, and electric fences that are used to protect seedlings.

There are also several commercially available tree shelters. These shelters are known for tall plastic tubes (usually made of polypropylene) and are used to preserve seedlings from animal browsing. In some cases, they may stimulate seedling growth by creating a "greenhouse effect" around the seedling.

The practice of tree shelters is somewhat a new method of seedling establishment and the long-term advantages of use which is yet to be determined.

One of the initial and primary obstacles faced with tree shelters was that songbirds like house wrens and eastern bluebirds fly inside of them (probably to feed on insects), but are unable to open their wings to fly out and then vanish within the tubes. As a consequence, it becomes necessary to cover the top of the tubes with bird-excluder nets.

Aside from commercially available tree shelters, some landowners create their seedling protectors from galvanized fencing. These protectors act as mini-fences and can be customized for single plants or groups of plants. All these methods for protecting plants who are on the verge of getting vanished or damaged by wildlife or livestock can be used on naturally occurring vegetation as well as newly planted stock.

6.3.6. Forest Edge Improvement

It is generally seen that majority of the species of wildlife make use of edge habitat for their several activities such as feeding, nesting, and traveling. The primary objective to engage in forest edge improvement is to enhance the overall food availability and cover along a forest edge by ensuring the continuous supply of a variety of vegetation types and layers, from the shortest herbaceous vegetation to the tallest trees.

The presence of multiple layers in the vegetation provides more places or areas where wildlife can feed and find nesting, resting, or escape cover. In addition to this, the recent research demonstrated that the nesting success of birds is higher along "complex" edges with multiple layers of vegetation in comparison with along "simple" edges with fewer layers of vegetation.

Edges between forests and fields can be used by both types of species, for example, the one that can be typically found within the forest and the one that typically found in fields. For instance, the black-capped chickadee, which is a forest species, is often seen to be nest along forest edges with the field sparrow, a species which usually exists in field habitat.

In addition, "edge specialists," like indigo bunting, are often found in the forest edges as it is their primary habitat. Other species such as the wild turkey, white-tailed deer or eastern cottontail may sometimes found to be feed along a forest edge as they pose the ability to quickly run away from the forest when there is any danger. Predators such as the long-tailed weasel or red fox often prefer forest edges as they can found a large number of prey there.

Various types of methods that can be adopted to improve forest edge habitat are discussed in this section, including planting, allowing the occurrence of natural succession, and cutting. Adding nest boxes and brush piles will also add habitat components along an edge.

The outcome of forest edge advancement work should be a wider edge habitat that provides a gradual transformation from the smaller vegetation in the adjacent habitat to the giant-size trees in the forest, while ensuring supply of food and cover for a number of wildlife species.

6.3.7. Herbaceous Forest Openings

Herbaceous forest openings refer to the openings in the forest canopy where the high amount of sunlight reaches the forest floor, providing support to the herbaceous vegetation. It is worth noticing that the herbaceous openings can be of diverse shape and size. They are majorly responsible for providing food, nest sites, as well as cover for specific species of wildlife. Fields, haul roads, orchards, utility rights-of-way, log landings, or openings formed within a forested area may all work as herbaceous openings, which means providing wildlife habitat. Wildlife species are often tending to get benefits from the herbaceous openings.

It is important to note that though herbaceous openings in a forest may provide ample to a variety of species, there are times when forming a forest opening can be detrimental to species which demands considerable unbroken expanses of forest (for instance, the northern goshawk, pileated woodpecker, some amphibians and neotropical migratory songbirds). Trade-offs between benefits and detriments should be vigilantly taken into consideration before creating new openings.

6.3.8. Nest Boxes and Other Nesting Structures

One of the proven ways to enhance habitat for wildlife is to provide nest boxes or structures. Nest boxes, platforms, and other types of nesting structures provide nest sites for wildlife in those areas where natural nest sites (especially cavities) are missing or available in very meagre numbers. It also helps in attracting wildlife to particular areas even when there are few nest sites.

6.3.9. Releasing Trees and Shrubs

A tree and shrub release refers to a technique or strategy which is used to encourage the growth of specific species, individuals, as well as groups of plants so that they produce more food or cover for wildlife. Releasing a plant carries activities such as removing other plants that are shading it and competing for sunlight.

Most releases are "crown releases." However, in some scenarios, the strategy of releasing roots from competition can also be taken into consideration. Crowns of selected species are often released from overhead shading on at least three sides to help enhance the growth.

When a shrub or tree release is being considered to improve wildlife habitat, the shrubs and trees chosen for release should be those that provide good quality food or cover for wildlife, such as fruiting shrubs.

The wildlife species that will get the maximum benefit from a shrub or tree release will rely on the wildlife species that exist on the property and the types of shrubs and trees were chosen for release. For instance, releases can be preplanned to enhance the production of acorn that will feed deer, squirrels, and turkey.

Releasing can also enlarge evergreen cover for mourning doves and ruffed grouse, or build the vertical structure in an understory that will help in providing foraging and nesting sites for the wood thrush as well as for other songbirds.

6.3.10. Releasing and Pruning Old Fruit Trees

Fruit trees are responsible for providing food to a large category of Pennsylvania's wildlife. For instance, in the fall, a white-tailed deer major portion of feed is apples.

Some other animals that take advantage of fruit trees include the songbirds, black bears, and small mammals. Also, as fruit decomposed on the earth, it provides healthy environment which is favorable for the existence of earthworms, that is why woodcock can often be seen feeding under fruit trees.

On some properties, old, decadent fruit trees can be present in young forests or abandoned fields. Old fruit trees provide clues to past land use. The property may have been a farm and these fruit trees were part of the backyard or an orchard of apple trees.

Over a period of time, as with the growth of other trees which shaded these fruit trees, there will be significant reduction in the production of fruit. Still, Fruit trees are often seeming feasible, and with a minimal focus can be shift to fruit-producing condition.

"Releasing" these trees may be essential, particularly if a young forest has grown over and around them. It is generally seen that fruit production largely depends on light. Removing other trees which will shade the fruit trees will help in eradicating competition for sunlight.

6.3.11. Spring Seep Management

Spring seeps refer to the natural water sources where fresh-water flows from below the ground to the surface to form small bodies or small streams of water. Spring seeps can be found in fields or forests, but are often present at the bases or along hillsides of mountains where groundwater flows to the surface. These areas usually have a small, year-round source of fresh-water. It is often seen that spring seeps provide ample benefits to a wildlife.

Spring seeps are especially vital during the winter when they may be the only source of fresh food and water. In the winter, groundwater is generally warmer (a consistent 50 to 55 degrees Fahrenheit) than ground and air temperatures.

Even during the harsh weather conditions, seeps usually remain unfrozen with flowing water and is favorable for green vegetation during the time when herbaceous vegetation occurs in small numbers. During harsh winter conditions, when water from other sources are frozen for longer periods of time, spring seeps are used heavily by wildlife.

In early spring, seeps areas are the first ones where vegetation grows. Thus, this food source provides food at those depressed time of year when the majority of the other food sources have been diminishing.

During the periods of deep snow, spring seeps also provide the support of snow-free travel lanes where wildlife can move and feed. Birds and mammals which take advantage of the herbaceous vegetation that persists and grows around seeps in the winter when other food sources exhausted.

Insects exist around the seeps ensure the consistent supply of high-protein food all over the year. Deer and small mammals find infinite sources of "browse" and other forage germinating around seeps. Bears and other berry eaters often get an advantage from the fruit-producing species that grow at an adequate level in moist conditions. Songbirds that get advantage from the fruit and insects around seeps, usually identify nest sites in the dense vegetation surrounding the seep.

Amphibians and reptiles that get the advantage from seeps which entails slow-moving water. Because most spring seeps do not create favorable conditions for fish populations, amphibian eggs can produce without heavy losses to fish predation.

Reptiles for instance turtles can get the advantage from the fresh spring water and soft mucky bottom of some seeps where they can bury themselves, allowing solace from hot weather conditions or remain dormant during the winter months. Reptiles and amphibians also get an advantage from the plant and insect food which exists around seeps.

One of the most critical management practices for spring seeps is to secure them from any kind of activities that could demean the seep, such as clear-cutting beside the agricultural or seep pollution. Some available ways for improving the habitat concerned with a spring seep include planting or releasing beneficial shrubs and trees around the seep and triggering the growth of herbaceous vegetation around the perimeter of the seep.

6.3.12. Stream Bank Fencing

It is generally seen that livestock which has free access to streams destroy fish and wildlife habitat, upsurge sedimentation and erosion, and degrade the quality of water. Stream bank fencing debar livestock from sensitive riparian areas.

After the exclusion of livestock, a buffer zone of vegetation germinates between the fence and the stream. This new riparian streamside vegetation plays very crucial role in providing food, cover, and nesting sites for birds as well as for small mammals.

Over 80 kinds of birds, encompassing herons, bluebirds, egrets, mallards, belted kingfishers, and pheasants, use streamside vegetation for summer nesting and feeding. Streamside vegetation enhances vegetation of fish by improving the quality of water, delivering protective cover, and enhancing food availability for fish. Stabilized stream crossings provide restricted areas where livestock are enabled with the access to the stream for drinking and/or crossing the stream.

6.3.13. Temporary Pools

Temporary pools refer to the wetland habitats that replenish with water at the time of the rainy season and then dry up later in the year. Many people regard temporary pools to be closely associated with vernal pools. "Vernal" means "of, relating to, or occurring in the spring," and these pools fill with water during the dry up in the late summer or in the spring and early fall.

However, there are also autumnal pools, that replenish with water in autumn. Temporary pools can be seen at a place where small swales collect runoff or intercept seasonally high-water tables.

Although it is worth noticing that the supply of water from these pools is temporary, it is vital because temporary pools are the hibernating and breeding grounds for amphibians such as spring peepers and red spotted newts. Temporary pools are not favorable for populations of fish so amphibian eggs can form without significant losses to fish predation.

These unique scenarios make temporary pools vital for the existence of many amphibian populations. Ecologically, amphibians are both predators that predate on small invertebrates and prey contributing to the existence of various other predators. Since the late 1970s, scientists across the world have been noted the consistent plunge in population of amphibians.

This fall makes conservation of temporary pool habitats highly important. In addition to providing hibernating and breeding habitat for amphibians, temporary pools also have the potential to provide support to a complex web of interactions between a number of organisms including salamanders, aquatic insects, frogs, snakes, turtles, large, and small mammals, songbirds, and waterfowl.

One of the best available ways with the landowners to provide temporary pools for wildlife is to prevent the present pools from destruction. Many temporary pools have been demolished at the time of development and other land-use changes. (Unfortunately, it is very difficult to identify temporary pools at the time of the dry season, and the majority of citizens all over the world don't realize the value of what looks to them like a mud hole.)

Temporary pools should be protected from deterioration due to the critical role that they perform within a larger habitat area. Usually, limiting any current degrading activities around a temporary pool is all that is essential to ensure or preserve the healthy environment of the pool.

Various initiatives have been undertaken in a way to create temporary pools for wildlife. These newly created temporary pools will help in offsetting some of the losses of natural temporary pools. However, the formation of a temporary pool demands special circumstances that do not prevail on all properties.

6.3.14. Warm-Season Grasses

Native warm-season grasses are prairie grasses that existed when ancestors of today's generation settled what was to become the Commonwealth of Pennsylvania. The four major grasses of tall-grass prairie habitat are big bluestem, switch grass, Indian grass and little bluestem.

Warm-season grasses, which are also known by the name "bunch grasses," grow in thick bunches rather than forming mats like many other grasses. These dense bunches of grass scattered all over the open spaces between the bunches which deliver valuable nesting and foraging cover for upland game birds such as pheasants and turkeys, several waterfowl, and other ground-nesting grassland species such as savannah sparrows.

Some warm-season grasses have the potential to grow in height as much as over 6 feet. Because of their height and rapid growth characteristics, these grasses provide tremendous cover for wildlife, especially waterfowl and upland game birds. Cottontail rabbits always take advantage of both the cover and forage formed by grasses of warm-season, and the majority of the songbirds feed on the small seeds of the plants. Undoubtedly, one of the key benefits of warm-season grasses is that, if these grasses were left uncut, they remain upright throughout the winter conditions, providing valuable cover for large number of animals at that depressing time of year when most other plants are dormant or died.

Planting warm-season grasses for wildlife has currently become widely used practice because of the high-quality habitat which is delivered by a stand of warm-season grasses. Planting warm-season grasses demands more patience rather than effort. After the plantation of grass, it may take from one to four years for a full stand of grasses to develop.

Because this habitat management practice has been recently gained momentum, new data or valuable information about to plant or grow these grasses has been revealed every day. Consequently, if one has interest in planting warm-season grasses for wildlife, it is recommended to take assistance from someone that has noteworthy experience in this field.

There are some of the widely popular methods being used; having god understating about them will give first-hand knowledge of what methods would be best according to the available factors and environment.

6.3.15. Wetland Restoration

There were some countries that have introduced the wetland restoration program. The primary objective to engage in the wetland restoration program is to improve the condition of already deteriorated land. In Pennsylvania, this initiative was undertaken by the U.S. Fish and Wildlife Service's Partners for Wildlife Program.

Wetland restorations are usually done on lands that were previously drained (in some cases to make cropland) and which are then restored to their original condition level as a wetland by eradicating the tiles or plugging the ditches that drain them.

The outcome of wetland restoration may dissimilar in shape or size and often have an area of open water with the burgeoning growth of wetland vegetation around the perimeter of the water. These wetlands play a very important role as these are responsible for nesting, breeding, and feeding habitat for waterfowl, amphibians, songbirds, and shorebirds. They are essential "stop-over," feeding and resting places for species which are migrated.

The excellent cover offered by the tremendous growth of vegetation around wetland habitats also provides support of surplus food and cover for species such as the muskrat, beaver, white-tailed deer, and cottontail rabbit.

Animals that greatly rely on the wetland habitats, and those animals that are generally considered upland species, often take the advantage from the restored wetland sites. It is generally seen that wetlands always help in reducing flooding and erosion, purification of water supplies by filtering sediments and pollutants out of the water.

6.3.16. Wildlife Corridors

A wildlife corridor is a habitat "patch" which is responsible for connecting two or more areas of undeveloped habitat that are far away or isolated from one another. There are chiefly two key types of wildlife corridors, but other areas can work like these two types on a very local scale.

The first type is a corridor which mainly have presence on a landscape scale. One good example of a landscape-scale corridor is a forested ridge top which links just like a "bridge" of wilderness, between two or more than two habitats which are far away from each other.

Corridors which have a presence on the landscape level are usually conceived to be working as a connective function, providing benefits to those species that demands large expanses of undeveloped habitat because of the fact that they have home which is far away from others, scattered over great distances, or need to travel to longer distance areas to find mates.

A second type of corridor generally has presence on a smaller scale, majorly on a local level, basically linking two far away habitats which are not primarily parted away by large distances. In agricultural areas, these types of corridors are mostly known by the name's hedgerows or fencerows. These "strip habitats" are responsible for providing food and cover for wildlife.

In case the hedgerow links two habitats, for instance, two woodlots on a farm, then it may be considered as a travel route between the woodlots by some species. Such kind of corridor can be connected to a property, providing potential travel passage and additional habitat for wildlife.

Another type of wildlife corridor which usually exists on a local scale is the buffer strip of vegetation alongside a river or stream that may differ in width. Chiefly rely on the length and width of the buffer strip, these areas perform the role of a travel corridors, apart from providing wildlife with good quality food and cover.

Connective corridors initiated by habitat management practices are in majority of the cases wooded patches of habitat which links two isolated woodlots or forests and will provide additional food and cover for wildlife.

It is generally seen that with changes in the land-use, there are results in the fragmentation of wildlife habitats, and habitat patches, for instance, more isolated and smaller forests. Perhaps one of the best ways to plant corridors is that the original landscape was interconnected.

There would be re-establishment of new corridors to some of those past connections. Corridors can also perform the role of a habitat patch, providing wildlife along with food and cover whether the animal lives in the corridor or just travels through.

The species of wildlife that will get the advantage from the wildlife corridors will primarily rely on the type of plants and other habitat components that have presence in the corridor, where it is situated and the category of species that exists in the adjacent habitat. Small mammals, for instance, the eastern cottontail and gray squirrel, and songbirds such as gray catbirds and song sparrows are some species that may adopt the use of wildlife corridors for nesting, feeding, or movement.

6.4. SUSTAINABLE WILDLIFE MANAGEMENT

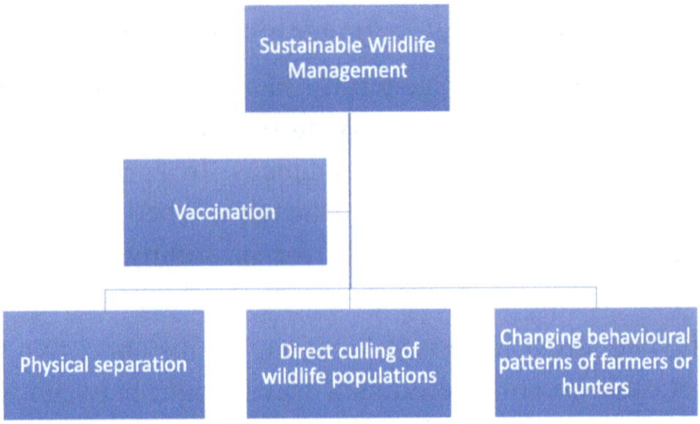

Figure 6.5. *Sustainable wildlife management.*

Methods that are used for managing outbreaks vary with reference to the type of disease, potential socioeconomic impacts and causes of transmission:

6.4.1. Vaccination

Vaccination to intercept any kind of disease is used in cases of highly devastating and pathogenic diseases, when there is a potential threat of transmission to pets or livestock, and to protect endangered or threatened species.

This method of disease control/ prevention was used between the year 2003 to 2004 outbreak of rabies in endangered Ethiopian Wolf populations in Ethiopia, and helps in impeding further mortality in their populations. In North America, some of the oral vaccinations such as raccoons, skunks, coyotes, wolves, and foxes have been used to control deadly disease rabies in wild populations near urban centers.

Vaccination against foot and mouth disease (FMD), as a strategic option, is best feasible for endangered or valuable species in the European Union, despite a region-wide ban on FMD vaccination as these are vulnerable to domestic animals because of international trade regulations.

6.4.2. Physical Separation

Figure 6.6. Cultural sustainability.

Source: Image by Wikipedia.

Physical separation, generally by fencing, has been successful in controlling and limiting transmission of some diseases between livestock and wildlife, but it should be planned vigilantly in a way to prevent effects on populations of wildlife through blocked migration passage or access to feeding areas or water.

In southern Africa, physical separation has been implemented successfully in order to prevent the spread of FMD from African Buffalo to live-stock by watchful separation of infected buffalo through artificial or natural barriers such as mountain ranges, rivers, and fences, in combination with vaccination of susceptible livestock living within buffer zones.

6.4.3. Direct Culling of Wildlife Populations

Figure 6.7. Sustainable agriculture.

Source: Image by Wikipedia.

The culling of wild species in a way to control disease is hardly used in SWM. Some alternate options can be use of sought, as removal is rarely effective, can interrupt habitats activities and cause dispersal of infected animals to new areas.

This type of method is majorly adopted in the UK to address the growing momentum of bovine tuberculosis on cattle farms, but some analysts are of the opinion that the limited culling of abundant badgers has aggravated the disease by increasing disease transmission and badger movement.

6.4.4. Changing Behavioral Patterns of Farmers or Hunters

Figure 6.8. The conservation atlas of tropical forest.

Source: Image by Flickr.

Livestock and wildlife owners, hunters, and farmers lay very important role in wildlife disease transmission and prevention. Improving health of livestock by restricting disease prevalence and parasites or insects that spread diseases will also help in minimizing the transmission risks to wildlife as well as livestock.

Some examples of wildlife-livestock transmission involves the introduction of HPAI H5N1 to poultry by waterfowl hunters (for instance, in Ukraine, Turkey, and Federation of Russia), and FMD from wild boar to livestock in Bulgaria as an outcome of hunting the boar and frequenting their locations of feeding.

It is important to raise awareness among hunters about health issues of animals that help in preventing pathogen transmission. Hunters can also make their contribution to surveillance programs for wild-life diseases by highlighting unusual events or by collecting samples which is of utmost importance to veterinary authorities as part of monitoring and surveillance.

6.5. PRACTICAL ASPECTS OF ENFORCEMENT

Figure 6.9. A team of geomorphologists monitors the glaciers and erosion rates in mount rainier national park, Washington.

Source: Image by National Park Service.

6.5.1. Enforcement Costs, Enforcement Budgets, and the Role of Fines

In addition to low fines, developing countries most likely have high enforcement costs. Beyond a conventional limit, the enforcement of access restrictions in developing country forests due to their large densely vegetated areas coupled with high pressure on their utilization is exorbitantly costly. Salaries for guards to patrol large areas along with vehicles, guard stations and other equipments are prerequisites for enforcement.

A considerable portion of many forest and park management budgets is formed by these costs. Guard and other enforcement costs make up the majority of the annual budgets in national parks in Thailand (MIDAS Agronomics Volume 3, 1993). About 60% of the forest department's budget in India is spent on enforcement activities that include salary and travel costs, and protection infrastructure (author's estimates).

The enforcement costs increase as the percentage of budgets when the prospect for tourism and recreational services is not very conducive. However, as anticipated in the literature and as seen in practice, there are the hindrances on the level of fines and those who have the supervision for the enforcement do not just reply on reducing the costs. Furthermore, the enforcement effectiveness is maintained by lowering the chances of getting caught which in turn reduces the detection cost and therefore comparatively increasing the fines.

Budgets available for conserving vital natural resources are slender or even zero in many developing countries. The need for large government provided enforcement budgets can be averted through policies that involve local communities in the management of the nearby resources through versions of Community-based forest management (CBFM) and Joint Forest Management (JFM).

But the access restrictions that are enforced by the villagers are required by these CBFM and JFM. Though this removes the cost for the government agency, it is considered as a social cost. For example, the place where JFM has been initiated in Tanzania, the responsibility for patrolling the forests is given to the village committees.

Furthermore, there is no budgetary provision for remunerating these activities (Robinson and Lokina, 2008). Undoubtedly the issue of the source of enforcement budget is largely overlooked whereas this is usually an important issue in a developing country framework.

With the availability of small or absence of centrally provided budgets, forest, and park management in developing countries faces a problematic situation. Alternatively, to boost the budget they base their reliance on cost recovery through the collection of hunting licenses, fines, and tourism gate fees (Robinson, 2008).

As a consequence of which enforcement efforts by park managers may be concentrated on more fruitful areas and offenders. Cost recovery tends to be achievable in places where there are charismatic species (Robinson, 2008).

For example, in Tanzania, 4.5 million US dollars were earned by the government in 1990 from 'trophy' hunting licenses in comparison to 1.9 million US dollars from the national parks system (Makombe, 1994). Likewise, no direct government funding is received by the parastatal Tanzania National Parks Authority (TANAPA), however, it has been able to collect the adequate funds from tourism gate revenues supplemented by external donors (Robinson, 2008).

The utilization of Hunting revenues to safeguard resources is not exclusive to developing countries. Presently, in the US, 'various licenses, fees, and taxes on hunting and hunting equipment fund more than 90% of the budgets of state fish and wildlife agencies' (Burnett, 2001). But in West Africa where there are few alluring species, there is no availability of these revenue streams.

6.5.2. Multiple Layers of Enforcement, Fairness, and Conflict

It is assumed by most of the economic enforcement literature that the sole optimizer in determining levels of enforcement is one actor or agency, who is primarily a private or government party.

The layers of jurisdiction or the incentives facing individuals in the implementation of an optimal plan are hardly integrated by the literature, with a few important exceptions. First, two tiers of enforcement, one identifies the illegal activities, and another lays down the rules and monitors detection agents were stated by Mookherjee and Png (1995). Other exceptions like Sutinen (1987) and Akella and Cannon (2004) involves employing an enforcement chain that perceives the importance of different individuals and organizations that are employed in different facets of enforcement such as detecting the crime, making the charge, finding individual guilty, and implementing the punishment.

Likewise, there has been an expansion from a single optimizer perspective in discussing how a group agrees to establish rules in some game theory and village institution literature. Despite these papers, the tiers of enforcement are commonly ignored by the enforcement literature. In reality, multiple tiers of enforcement are not an exception but are instead regarded as a standard.

For example, there are three tiers of detective agents who may have divergent incentives in enforcement in Orissa at the Similipal Biosphere Reserve (SBR). The first tier has the forest guards who are low paid and are the ones who patrol the area with minimal resources and administration usually using bicycles or on foot. While the second tier is of the foresters who will oversee the first tier of guards, and along with this they also assume various activities in enforcement.

Also, for every 8–10 beats, there will be one forester in-charge of. The third tier is made up of forest managers and rangers, who have the authority for the forest protection strategy development and its supervision. This tier is assumed, in theoretical models, to be the individual for the optimization of the forest. It is to be noted that, in the three tiers there is differentiating motivation levels and level of influence.

However, as the guards of the third tier are the topmost and in charge of the enforcement, they are able to control the type of enforcement and the extent of it. They have so much influence that they have the choice to take bribes, how much finances and efforts should be used for enforcement and who to punish for crimes in the area and the officers, and whom to not take action against.

Such decisions taking by the top tiers' officers are unnoticeable to the other tiers in the enforcement. These decisions are generally biased towards who to formally or informally punish and from where the bribes should be taken. If we look at Orissa, many of the locals who are the foresters and guards consider some actions as "more illegal" than illegal activities and this is incorporated into their decisions.

These guards often ignore the illegitimate activities in the forest reserves or national parks by the locals, as they are considered to the reality that the new laws which govern the PAs have been a disadvantage to the villagers for their traditional access to the forest. As previously discussed, guards choose who to punish, which gets shaped by the relative strength of the pockets of the illegal extractor which then creates dissatisfaction in the community who feel they will be punished as compared to others.

As a consequence of the informal decisions made by the guards, concerning the acceptance or "fair "level of extraction and punishment, the enforcement regime is close to achieving social optimization which considers both the resources and livelihood. However, due to the lack of biased decisions, this may do the opposite and move away from the social optimum.

Developing countries face issues of fairness where poor villagers, who may have traditionally though illegally relied on resources from a particular protected area as a key contribution to their livelihoods, undertake the illegal collection of resources.

Concept of fairness is dependent on a lot of factors such as the count of villagers who appear to be 'getting away' with the illegal activity, the purpose of collection of resources i.e., for home consumption or commercial selling and whether the caught individual is the local of that area or village. According to (Sutinen and Kuperan, 1999), if the rules are recognized as being fair, they tend to be accepted.

The tendency of guards to make differentiated enforcement decisions based on their concepts of equity and compassion especially with reference to extraction for subsistence home use are reflected in our interviews in Tanzania, China, and India and one author's experience as a forester. It's mostly assumed that there's a fixed relationship between the amount spent on enforcement and its effectiveness by the optimal enforcement literature.

Though there are exceptions like Mookherjee and Png, 1995). However, in reality, those who are actually responsible for enforcement chose the amount of effort to put into enforcement such as the forest guards in Orissa, or the village committee in PFM in Tanzania.

With a lesser level of monitoring along performance and rewards not being in sync, the guards do not give much effort as they are not provided with many incentives. Therefore, there is the freedom to determine who the guards can book someone for an offense as there is not a high degree of supervision.

For instance, In India, guards are not truly booking offenses but rather are turning the eye. In comparison, when someone is caught for killing a rare animal or with stealing timber, guards will go with the easier option of recording the timber crime where the individual just has to pay a fine and leaves with the timber because this offense requires fewer efforts and paperwork.

In this example, if the offense is properly booked, the guards will have to make a genuine entry into the record books, and then move the illegally obtained firewood to a safe storage unit it is disposed of, and additionally, these guards may be present in court during the trial; however, due to this excessive work, the guards prefer to not book the major offenses.

6.6. CONCLUSION

In the end, it is concluded that the management and use of wild animals give rise to ethical dilemmas and disagreements in which human preferences, needs, wants, and interests, concern for individual animal welfare, and the value of ecosystems, biodiversity, and wild nature are part of the discussion. The formation of policy dependent on the way in which these diverse values are prioritized. One cannot make any particular recommendation.

All these are varying on various factors and local conditions. However, it is also not suggested that explicit consideration of the values at stake should underpin careful debate about, for instance, whether continuous human activities involvement in nature reserves and other wild areas is desirable, and what constitutes "good" and "bad" human interventions in relation to wildlife.

REFERENCES

1. *Ethics of Wildlife Management and Conservation: What Should We Try to Protect? | Learn Science at Scitable*. (2020). [online] Available at: https://www.nature.com/scitable/knowledge/library/ethics-of-wildlife-management-and-conservation-what-80060473/ (accessed on 21 August 2020).
2. Penn State Extension. (2016). *Management Practices for Enhancing Wildlife Habitat*. [online] Available at: https://extension.psu.edu/management-practices-for-enhancing-wildlife-habitat (accessed on 21 August 2020).
3. *Re-Conceptualization of Wildlife Conservation*. [ebook] Available at: https://extension.psu.edu/management-practices-for-enhancing-wildlife-habitat (accessed on 21 August 2020).
4. Robinson, E., Kumar, A., & Albers, H., (2010). Protecting developing countries' forests: Enforcement in theory and practice. *Journal of Natural Resources Policy Research*, *2*(1), 25–38.
5. *Sustainable Wildlife Management CPW and Animal Health*. (2015). [ebook] Available at: http://www.fao.org/3/a-i4891e.pdf (accessed on 21 August 2020).
6. *Wildlife Management Practices (WMPS)*. [ebook] Available at: https://extension.tennessee.edu/Putnam/Documents/WMPs%202016%20WHEP%20manual.pdf (accessed on 21 August 2020).

7

Impact of Conservatory Practices

CONTENTS

Conservatory practices are important for the conservation of wildlife. A variety of conservatory practices are being used these days in almost every nation to protect their respective wildlife and thus, biodiversity. However, use of different conservatory methods or practices has some major impacts on the environment and societies also.

The chapter initially discusses the basic importance of wildlife and conservatory practices. Later, some social impacts of conservation and projects regarding wildlife conservation are explained. In detail, the impacts of nature conservation intervention on the well-being are well explained. Certain key impacts of conservatory practices on food security, human health and creation of opportunities are discussed in brief.

7.1. INTRODUCTION

Similar to forests, wildlife is also one of a national resource. Wildlife does not just support in the maintenance of the ecological balance, but it is also valuable from an economic, aesthetic, and recreational perspective. There was a point of time when human interference was much less to the number of wild animals, that were quite high. Also, at that time there were no issues related to their conservation and protection.

Although, with the development of evolvement of agricultural practices, settlements, and various other industrial developments (specifically because of the greed of human beings), the number of wild animals progressively decreased. It eventually resulted into the extinction of a lot of many species, while a lot of many are now on the verge of extinction.

Deforestation (the process of cutting down trees for the use of that land for different purposes) is one of the key reasons for the major loss of wildlife. All around the world, mass killings of several wild animals for their meat, fur, bones, teeth, skin, hair, etc., are going on. As a result, the need for wildlife conservation has become a very necessary part now.

Growth in population, increase in agricultural practices, construction of roads and cities and the increase in pollution are amongst the several pressures that are on the natural habitat of wildlife. Together with illegal hunting, reduction in natural habitat and its degradation has exposed the biodiversity of various regions where these are present in an extensive amount.

Conservation of preservation of wildlife does not necessarily mean a kind of blanket protection to all the floral and faunal species. Instead of this, it requires proper judicious control over the multiplication of plant and animal species, that interact with each other in order to provide an optimal environment to human beings whose existence at present is in danger.

Because of the irrational use of natural as well as biotic resources of the earth in the past, a lot many wildlife resources have been threatened beyond recovery. Now, it is an urgent responsibility of humans to protect the natural glory of ecosystems and also to develop a system of co-existence with almost every living creature on the planet.

However, most of the nations of the world are much specific in regards to wildlife conservation as the number of wildlife animals is decreasing day by day. World Wild Life Fund is an international agency. This agency is doing a very commendable work to promote the protection of wildlife. Some other national agencies that are also involved in the conservation or preservation of wildlife.

Some Steps in the Direction of Wildlife Conservation Could Be as Follows:

a) To do survey and collect all the valuable information related to wildlife, mainly their number as well as growth;
b) To conserve habitat by conserving forests;
c) To restrict the areas of all the natural habitat;
d) To protect the wildlife population from natural hazards and different types of pollution;
e) To enforce complete restriction on the activities related to hunting as well as capturing of wildlife;
f) To enforce strict restrictions on the import and export of wildlife products;
g) To give a very harsh punishment to those who are found to be involved in such activities of import and export;
h) To create game sanctuaries for particular wild animals;
i) To make accurate and proper arrangements for the protection of those species that are present in very number or those that are on the verge of extinction;
j) To develop general awareness programs at the national as well as international level about the conservation of wildlife; and
k) To adopt a kind of system of wildlife management by trained professionals.

To better understand the conservation of wildlife, India is a good example to study. In India, numerous steps have been taken for the protection of wildlife. This country is known as a country of diverse wildlife. There more than 500 different types of wild animals, 2100 different types of birds and around 20,000 different types of fishes and reptiles have been found.

According to an estimate, around 200 species of wild animals and birds in India have already extinct. On the other hand, almost 2500 other species are found to be on the verge of extinction. Some of them are chinkara, blackbuck, swamp deer, wolf, nilgai, tiger, antelope, girl lion, rhinoceros, crocodile, pelican, flamingo, white crane, bustard, mountain quail, grey heron, etc.

In India, the NGOs and the government are taking much interest in the conservation of wildlife. The Wild Life Protection Act, 1972 has various provisions for wildlife conservation. Almost 165 game sanctuaries and almost 21 national parks have been developed for the protection of the natural habitat as well as wildlife animals. Along with all these things, a Wild Life Conservation Week is also celebrated in October every year. However, still there is a long way to go in this particular direction.

Wildlife mainly includes all of the undomesticated living plants as well as animal species. All of the living organisms, such as humans based on other living things for survival. The advantages that wild species will give to people are the eventual outcome of the proper care as well as the management of the wild species by humans. By the management of wildlife, humans are actually preserving the environment.

While conserving the environment, human beings are maintaining a very healthy and prosperous world in which they can live happily. Here, the question arises who is mainly responsible for the management of wildlife? A responsible wildlife management must be a goal of every landowner, each, and every outdoor enthusiast and of course, every single person.

Majority of land in the United States is private ownership. For example, in Nebraska, private ownership accounts for a total of 95% of all real estate ownership. From this, it can be observed that the wildlife refuges cannot be accountable for all the wildlife management activities alone.

Next question is, where is the motivation or encouragement for the individuals to presume the role of wildlife managers? It is not very difficult to recognize the value of domestic plants and animals. But, how can be the worth of wild species measured? Wildlife management gives humans much more than a healthy environment.

There are four major categories in which the wild species of plants and animals offer values to the existence of human beings. Recognizing each of such areas will eventually develop an understanding of the significance of wildlife along with its management.

In general, conservation of wildlife has two meanings. One of them is the preservation of species as well as species diversity. While, another meaning depends largely upon the welfare of animals that is basically aimed at wildlife in captivity (Lu, 2009).

Education about conservation is very critical component of environmental education. It is mainly aimed at increasing awareness in humans about the conservation of biodiversity and also at the altering environmental altitudes as well as behaviors in order to encourage conservation by giving education and practical activities.

Education about conservation of wildlife forms a key part of environmental education. As environmental concerns have been observed to be increased all across the societies, conservation of wildlife has now become a very important social issue. Although, there are some considerable differences in the notions of wildlife conservation, along with various ideas of reasonable protection that are currently being debated.

Some people believe that conservation of wildlife must be incorporated with the protection of animals and should follow the principals of animal protection and conservation. In addition to this, such kind of protection comprises of resistance to live bear bile that is utilized in conventional Chinese medicine and ban on hunting and some other related activities.

In addition, an absolute conservation has very strong public sensibilities that can eventually result in extreme activities related to conservation of wildlife, hence animal and welfare rights. Comparatively, some other individuals believe that the conservation of wildlife must be completely based upon the scientific strategies and attitudes.

Absolute protection, unfortunately, dominates the public opinion as well as sympathy at present (Zhang et al., 2015), together with rational thinking and scientific discussion often discounted. Absolute protection mainly comprises the random release of animals. This has eventually led to the assault of alien species that are playing a negative role in hindering and weakening the process of wildlife conservation (Karanth et al., 2008).

As attitude of public plays a major role in the formulation as well as implementation of the wildlife conservation management policies, some of the key issues that require to be solved in the construction of an ecologically-based civilization are explaining wildlife conservation, encouraging scientific wildlife protection and promoting an objective view of the wildlife conservation by the public.

7.1.1. Wildlife as an Important Resource in Human Evolution

Research has shown that Homo sapiens (archaic) and Homo habilis increased the amount of meat protein in their diets by the activity of hunting, which eventually result in rapid development as well as increased volume of the brain (Wang, 2004). Furthermore, meat was important for both the process of reproduction and evolution of prehistoric humans (Niche, 1995).

As compared to Homo habilis, Homo sapiens (archaic) made use of fire to cook food and this in turn resulted in major changes in human morphology (Qiao, 2011). An effective decrease in the chewing time of food, as well as digestion had resulted in critical changes in the mandible as well as dental morphology like smaller teeth, smaller crinium visceral, jaw retraction and rostrum retraction (Yu and Zhai, 2004).

Additionally, human beings started learning the use of fire and the beneficial consumption of meat protein, which is easily absorbed as compared to the raw meat, following the use of fire in cooking encouraged the development of the human brain as well as body.

In addition, activities of hunting improved human cognition along with their problem-solving abilities and also endorsed the evolution of physical fitness (Qiao, 2011). The consumption of meat promoted the health of humans by increasing the resistance for disease and establishment of the functions of the brain and some other organs (Qiao, 2011).

7.1.2. Importance of Wildlife in the Long-Term Development of Human Society

Wildlife has played a significant role in the development of medical materials, clothing, scientific research and experimental models. Furs and skin of animals have been utilized for clothing, even today, these are used as the symbols of fashion.

Also, it has been found out that wild animals provided nutrition to humans and formed an essential part of diet. The farming and domestication of wildlife, the development of feed technology and the invention of meat as well as milk production can be regarded as the three key revolutions of the human diet.

All these developments improved the proportion of fat and protein in the diet and improved the development of the human brain and body. At the initial stage of Homo sapiens, wide-spread farming practices and the initial formation of agricultural production increased the variety and abundance of human food. Again, this contributed to increasing the volume of the brain.

At the later stage of Homo sapiens, with the introduction of primary agriculture, animal husbandry and the industrial revolution, the habit, structure, and the notion of the human diet has tended towards much stability. In a corresponding way, the physical form has remained unchanged predominantly (Qiao, 2011).

Also, medical advancement has been found to be very important in the development of humans. Animals form a major part of the traditional Chinese medicine, as raw materials as well as secretions. It also plays a crucial role in the continuation of Chinese descendants (Liu, 2014).

Wildlife, in addition, provides raw materials for medical and scientific research. Wildlife and humans are highly interrelated to each other in terms of culture, food, hunting, and eco-tourism also. Hence, it is quite difficult as well as inappropriate to separate wildlife from human.

Therefore, it is socially as well as scientifically important that people clarify how best to conserve the wildlife and upon which the notion of wildlife conservation is based.

7.2. ADDRESSING THE SOCIAL IMPACTS OF CONSERVATION

As a social process that involves the decision related to access, use, and sustainability of the natural system of the earth, the process of conservation inevitably comprises both the social benefits and the social costs. Benefits mainly consist of the clean water and air, the growth and survival of nature-based economies and the approaches of life, along with the esthetic and cultural values of the wildlife.

Costs in general consists of restrictions on resource use for the social changes and economic purposes linked with altered patterns of the resource use. Distribution of the costs as well as benefits of the conservation, differs greatly across the strategies of conservation.

In the case of protected areas (PAs), there is an increase in recognition that a lot many costs of the PAs are locally borne, mainly by the poor communities, while, on the other hand, the benefits accrue at a global scale (IUCN, 2005).

Local costs that are mainly linked with the structural forms of the PAs can comprise limitation on the use of natural resources, physical displacement, limitation on the access for cultural and religious purposes, human-wildlife conflicts and conflicts arising from the enforcement activities.

Concerns related to the social impacts of conservation-both positive that is benefits and negative that is costs is not new. It has developed as a major part of the key concerns in terms of social justice in case of conservation policy since the year 1970 and in practice since the year 1980 by approaches like integrated conservation and the development as well as community-based natural resource management (Fortwangler, 2003; Adams and Hutton, 2007).

At the same time, certain concerns regarding the social impacts of the process of conservation have not yet been resolved completely. While in some other aspects, it is increasing. As public PAs have been a key point of debate regarding social impacts, the issue is much relevant across a broad range of conservation strategies.

The chief aim of this section of the chapter is to review some of the past experiences from the point of view that how they have sought to deal with social impacts, to recognize the limitations as well as lessons that can actually inform the future practices.

It can be contended that with key recommendations on how conservation organizations can effectively contribute to ensuring that social impacts are more effective and equitable addressed, along with in-depth social analysis, increased support for the community-based approaches, organizational policies and policy-oriented partnerships.

7.2.1. Why a Concern with Social Impacts?

Concern for social impacts in the field of conservation has ethical as well as practical foundations (Wilshusen et al., 2003; McShane, 2003). Ethical foundations arise from the recognition of rights of the local people to their specific resources and also as a means of subsistence and for the compensation of losses.

Such kind of ethical dimensions rest largely on the social justice values and the legal human rights or frameworks. In addition, these are necessarily associated with the outcomes of conservation. Practical foundations stem from the connections between conservation and social outcomes.

Such kind of connections basically operate in a great number of ways. For instance, a flow of positive advantages to the local people from the natural resources can effectively enable and offer incentives for the sustainable management over a long period of time. On the other hand, social impacts that are negative can erode the local support along with global constituencies for the process conservation, making conservation less sustainable and more difficult to practice.

Usually, social impacts and conservation are associated by the wider forces that harm the biodiversity as well as the well-being of the local people like commercial over-extraction of natural resources. Major attention to such linked social impacts as a part of the conservation initiatives offers a basic foundation for alliances with certain social groups in order to address the common concerns.

7.2.2. Experience and Lessons

Several efforts have been made to address the social impacts of conservation practices. These efforts have spanned a broad range of approaches such as integrated development and conservation, targeted compensation and community conservation. One of the effective approaches has been to associate the management of state PAs with the activities of local community development.

This specific approach is mainly associated with an integrated conservation of wildlife resources and development projects that were executed in the 1980s and 1990s as a chief strategy for conservation practices, however it remains highly relevant. All kinds of development activities have taken different forms, specifically economic alternatives to use resources in the protected regions, however social services also (e.g., health, and education).

In certain cases, revenue sharing from the fees of parks and some other sources has given additional resources so as to support the services of the community or the establishment of alternative activities that can be used for the generation of income.

Major aims of such community development activities have been always to decrease the human pressures on biodiversity and also to generate the local benefits such as, as a form of compensation for limitations on resource access that were established by the PAs (Larson et al., 1998; Brown and Wyckoff-Baird, 1992).

To address the social costs, the development activities were linked to the state protected area management and these were having various limitations. One of these limitations in their association with the exclusions of the protected area, alternative livelihoods and some concerned activities that have tended to take the compensation of social costs, instead of the preservation as the beginning point.

Simultaneously, a growing body of research has questioned the rationale in terms of ecology for the exclusionary approaches in order to protect the area management.

Analyses of the conservation discourse trace the determined images of 'wilderness' all over the history of conservation. It also contends that kinked assumptions of a principal incompatibility of people, as well as wildlife, have driven certain actions in order to separate people from nature in specific places (Colchester, 2004; Adams and Hutton, 2007).

Certain other critiques driving from the conservation practices have accentuated the issues or problems of flawed or not enough social analysis in the project design (GEF, 2006; Seymour, 2008), specifically the analysis of the actual level of the conflicts between the objectives of conservation and the patterns of the local use of resource or the viability as well as the appropriateness of the alternatives.

One key aspect of the flawed social analysis has been observed to be the inclination towards focusing on the local issues and their effective solutions (Larson et al., 1998), as such kind of risks exaggerate the influences of local use on the biodiversity, and confusing external factors and broader drivers.

An overall inference of such critiques is that, even where the key aim of the alternative livelihoods and some other development activities has been specifically to produce benefits for the local communities. While the approach does not actually challenge the assumptions related to the incompatibility of individuals and the nature that eventually give rise to exclusions of protected area along with their associated social costs.

Analysis of some alternate livelihood activities as a kind of compensation shows an additional set of restrictions. One of the limitations is that the association between costs and benefits has usually been vague; without tangible evaluations of nature as well as the distribution of the impacts of certain PAs. There are quite less chances to be directed towards the most affected people.

The sequencing of activities related to protection and development has also tended to be de-linked. Due to increased limitations on the access and utilization of all-natural resources can be put in place very quickly as compared to benefits, mainly from enterprise-based development activities. Advantages or benefits have not started flowing until very long after costs have been acquired.

Where the flows of benefit have been not realized or limited, there the aim of compensation has not been attained. Developmental activities and alternative livelihood also tend to concentrate on the economic costs along with very little attention to maintain the ties between community and culture, knowledge, governance systems, and certain other social values associated with resources and land.

Eventually, as the objective of alternative income-generating activities or social services is to produce social benefits, liability to communities for such benefits-linked to articulated and defined costs have tended to be restricted.

Making a contrast to such experience is a variety of place-based or community approaches, for example, community forestry and fisheries, indigenous territorial management and community-based wildlife management.

All these approaches have usually considered as an initial point of the connectedness of people with their resources and lands and acknowledgment of human use as a key part of landscapes and ecologies. As observed in context to social impacts, community-based conservation (CBC) and natural resource management (CBNRM) have sought to prevent or avoid the social costs in general. It is done mainly from the effective limitations on the access and use of the natural resources and also to generate benefits from the management of communities.

As all such initiatives have been most often focused on the PAs, the community conserved areas are now increasingly being identified as a type of governance, together with the state and private management for the PAs (Borrini-Feyerabend et al., 2004).

Initiatives that are based on the communities have significantly demonstrated some major successes in producing positive outcomes related to conservation, at large scales. Some examples comprise community forestry in Mexico, Namibia's National CBNRM program, and locally managed marine areas in the Pacific.

Although, some considerable gaps are there between the objectives of CBNRM and community conservation and their implementation in real. Critiques from the social point of view focus towards to still-too-limited level to which the authority and power have been abandoned by the government to local and indigenous communities. For instance, by the recognition of rights to resources and lands, conferring of authority in the customary governance institutions, and certain benefits produced from the use of natural resources along with equitable distribution of revenues (COICA, 2004; Jones, 2004).

In this regard, social costs can be considered as an outcome of people doing investments in the management of activities that are not completely compensated because of the generation of very limited benefits and its distribution to the communities.

Another major approach comprises 'targeted compensation' efforts that are basically used to address the issues or conflicts between objectives of conservation and different aspects of human use. Some examples consist of direct compensation to forgo particular kinds of harvesting, hunting, or cultivation activities like under planting and grazing in forests, that affect the biodiversity or payments for the attacks by predator species of high conservation values on livestock.

For instance, in case of Namibia, the government offers an essential compensation to all the community conservancy members for predations on the livestock by wildlife (WWF, 2008b). The major elements of the promising efforts that are made in order to establish such types of the targeted compensation are that the objective of specific conservation as well as the social impact is defined in a very clear way, the kind and the level of compensation is agreed after being negotiated, and the accountability for compensation becomes clear.

7.3. WILDLIFE CONSERVATION PROJECTS DO MORE HARM THAN GOOD

Eco-tourism and the conservation projects of western-style are harming the wildlife to a greater extent. It is causing damage to the environment and displacing as well as criminalizing the local people also.

The wildlife tours and pristine beaches that are demanded by the overseas tourists had ultimately led to such kind of development that do not even benefit wildlife, for example, beaches being made, waterholes drilled, mangroves striped out, and forests cleared as per one of a world expert on the ethical dimensions of wildlife conservation as well as management namely 'Rosaleen Duffy.'

All these things usually hide a 'darker history.' Nature Crime: How We're Getting Conservation Wrong, a new book of Rosaleen Duffy that draws on almost 15 years of research, around 300 interviews with the conservation professionals, tour operators, local communities and the government officials was published to make people better understand about wildlife conservation and how people are taking wrong or not worth efforts to conserve wildlife.

Duffy added, when wildlife reserves were added, the local communities can instantly find that their regular subsistence activities like hunting and collection of wood, have been banned.

Figure 7.1. The conservatory practices involve the communities in them and thus prevent any local harm to the animals.

Source: Image by Flickr.

Simultaneously, well-intentioned efforts or attempts in order to protect or conserve the habitats of animal species on the edge of the level of extinction eventually lead to the development of wild and people-free areas. Such kind of approach has led to the displacement of people living all around the world.

As claimed by Duffy, 'professor of international politics at Manchester University, 'the process of conservation does not comprise neat win-win scenarios. Distinct kinds of schemes come up with certain rules and regulations that criminalize the communities, turned out to in the language of partnership as well as participation, associated with promises of employment opportunities in the tourism sector.'

The major failure of the western-style conservation method is the presumption in which people or humans are considered to be the enemies of wildlife conservation. They are generally regarded as the illegal traders, the hunters, the poachers, and the habitat destroyers.

It is believed that people who were involved in the process of conservation are known as 'wildlife saviors. Well, these things can be called as oversimplifications. The inability in negotiating conflicts related to wildlife conservation and working with people on the ground is where the approach of conservation sows the seeds of its own doom.

The question arises why some people who try to conserve the wildlife end up opposing the local communities against the conservationists? It is mainly because they are considered as unjust impositions in spite of their good intentions. This is important because failing to deal with injustices eventually damages the wildlife conservation in the long run.

All these things should not stop people who are supporting the schemes of conservation. Wildlife is under threat and every responsible person is required to act urgently. People should be encouraged by the environmentalists to identify what are the real costs as well as benefits of conservation, so that much better practices can be developed for people as well as animals.

Some of the assumptions in which the ends justify the mean results in a specific situation where the global conservation movement along with their supporters all around the globe presume that they are making ethical as well as environmentally sound decisions to save wildlife. In general, they are supporting the practices that have unethical, counterproductive, and completely unjust outcomes.

It is the deceptive belief that ecotourism is a key solution to the problem of promoting economic development in a very environmentally sustainable way. This is a 'bewitchingly simple argument.' However, the assumption that such kind of tourism essentially translates into the types of development that benefits wildlife is very much simplistic.

It has been found that the holiday makers are not aware about how their tourist paradises have been produced, in most cases. They generally assume that the silver Caribbean beach and the picture-perfect landscape is a natural feature. Though, this is really very far from the truth.

In a similar manner, construction of hotel in some of the tropical regions can eventually result in naturally important mangroves or beach buildings which damages the coral reefs. The World Wildlife Fund (WWF) for nature which is amongst the four biggest environmental NGOs in the whole world, sustains that the loss of wildlife is a key challenge that our planet is facing.

Intrinsically, a very powerful focus on wildlife conservation is important. According to a WWF-UK spokesman, "conservation is a necessity so let's not throw the baby out with the bathwater." "There are certain examples where ecotourism was working well and has thrown a complete lifeline to the communities in context to social benefits and economies, along with added benefits of biodiversity.

He further added, "Let's have more of those projects that are working for everybody and everything." "There is no one-size-fits-all when it comes to ecotourism and conservation."

7.4. IMPACTS OF NATURE CONSERVATION INTERVENTIONS ON HUMAN WELL-BEING

The world conservation strategy that was published by the International Union for Conservation of Nature (IUCN), the United Nations Environment Programme (UNEP) and WWF (WWF) in the year 1980 was the first major international policy program to make the associations between human development and conservation of nature very powerful.

All over the phase of the 1980s and 1990s, conservation, and developmental organizations encourage and then implemented a great number of initiatives that were basically aimed at improving the advantages of nature conservation for the population at local as well as wider scales.

All these have taken the form of integrated conservation and development projects (ICDPs), community conservation, community-based natural resource management and so on. The phase of the 1980s observed a key increase in focus on human rights mainly in the movement of conservation, specifically on indigenous rights as well as community displacement from all PAs.

During the initial phase of 1975, the IUCN General Assembly adopted a specific 'recommendation' that the rights of indigenous people must be considered in national parks and various other PAs. In the year 1982, this was reaffirmed at the World Parks Congress and subsequent IUCN meetings.

Such attention to the human dimensions of conservation has significantly increased since the turn of the new millennium. A broad range of reports like The Economics of Ecosystems and Biodiversity and the Millennium Ecosystem Assessment has tried to make people knowledgeable about the association between the well-being of humans and nature.

On the other hand, some of the international policies such as the UN Conference on Sustainable Development Rio+20, the Millennium Development Goals and their successors and the Convention on Biological Diversity have focused much on how sustainable natural ecosystems are associated with the benefits for the well-being of humans.

These kinds of associations consist of essential services like fisheries, purification of water, and pollution that support the good health of humans, support livelihoods and contribute to food security.

Along with the policy shifts, a great number of major international conservation organizations such as Conservation International (CT), Birdlife International, Fauna, and Flora International and The Nature Conservancy clearly reference people in their vision and mission statements and aim to attain some of the key outcomes that are socially beneficial from their conservation efforts.

CT explicitly revised its mission in the year 2007 so as to support the well-being of humans by restoring as well as protecting the ecosystems that offer essential goods and services. In the recent past, guidance to address targets of human well-being and social results and outcomes in all of the conservation projects has been developed.

This kind of evolution towards a much greater focus on the human dimensions of the efforts related to conservation has taken place for a variety of reasons. Over the past few years, the process of conservation has been represented as a win-win solution for the alleviation as well as sustainable development, and also as a limitation on economic growth.

As a lot many conservation projects have attained conservation and development goals both, some conflicts along with negative relationships between human well-being and conservation have also been accentuated, comprising human and wildlife conflict, loss of access rights, and evictions from the PAs.

Hence, the conservation policy has been influenced by a major requirement of demonstrating contributions to broad development goals, for example, USAID's Biodiversity Policy (2014), together with a real desire of securing and ensuring the endurance of natural ecosystems upon which the vulnerable populations actually depends.

Basically, this gradual increase in focus towards the human dimensions of the conservation process, a lot many predominant hypotheses are there regarding the effects of conservation interventions on the tangible as well as intangible domains of human well-being. The tangible domain mainly comprises economic and material wealth, physical, and physiological health, etc. On the other hand, the intangible domain comprise culture, social cohesion, good governance and equity.

These tangible and intangible domains of human well-being are usually interdependent as well as mutually reinforcing. For instance, the presence of tangible benefits of well-being like securing jobs, and securing good health of humans can encourage well-being of households as well as the community. Whereas, good governance can eventually result in access and security of livelihood resources that actually improve the well-being of the economy.

As various indices have been developed to measure and document different domains of human well-being related to conservation interventions. These have not been linked with the tools and approaches for evaluating the strength of such relationships, nor the outcomes of these outputs on the management decisions.

Along with an increase in focus on the human dimensions of conservation, certain decisions are important related to what kind of interventions are significant in meeting the social as well as biophysical objectives, which of the indicators are correct to use, when, and how changes are expected to take place and what kind of evidence are available in order to support such decisions.

Till now, the state of current information or evidence from the rigorous and relevant research, noting down of the impacts of nature conservation on the well-being of humans are not clear. For instance, as it is largely been identifies in literature and emergent policies that the ecosystem services are a key part of the environmental provision for the well-being of humans. There are still a great number of questions that remain around the particular role they play and the casual relationships between services and their outcomes.

For example, as demonstrated for the direct payment schemes, the conservation interventions may impact the well-being of humans by direct provision of the income from payments that allow a much better livelihood, though also in directing by ecosystem services.

Payments might encourage advocacy for the conservation of natural ecosystems that eventually lead to benefits of human well-being, for example, nature-based livelihoods and sustaining sources of wild food. Unintended, negative consequences might also evolve out from the flow of novel income sources into the communities along with the issues of equity.

A lot many issues affect the capability of the conservation sector to assess the contributions adequately to the well-being of humans. First of all, the interactions as well as dependencies between social and natural systems have yet to be integrated into the current assessment frameworks, that are completely focused on socio-economic and ecological benefits.

Empirical evidence that demonstrates the association between the well-being of humans and healthy ecosystems is still lacking, along with reporting and assessment of the conservation outcomes that are inconsistent and limited.

As a result, in the lack of a coherent evidence base regarding what actually works and what does not, the conservation organizations together with their partners are trying to implement certain interventions under assumptions in context to likely effects.

Such uncertainty brings a considerable risk, as the investments could fail to attain any one or both the social and ecological objectives, waste scarce available resources, and harm the credibility efforts with the decision makers, donors, and certain other stakeholders.

Some of the recent studies have started characterizing the evidence on the relationships between humans and nature. Such kinds of efforts are related to definite kinds of conservation interventions such as community forest management, PAs, etc., payments for the ecosystem services, alternative livelihood schemes, eco-certification, and specific components of biodiversity.

Work has not been mainly done to articulate frameworks for the analysis of impacts and the casual pathways of certain conservation projects on the social as well as natural outcomes. Recent advances in the conservation approaches are encouraging.

However, the existing studies have taken a narrow focus to examine the links between well—being of humans and conservation, observing one particular kind of intervention on a particular kind of outcome that limit the insights into interdependent contexts as well as outcomes along with the wider landscape on which the outcomes of conservation play out.

Several studies and researches have been conducted to identify the impact of different kinds of interventions upon a broad variety of potential outcomes and the casual pathways. Therefore, the benefits of larger review consist of:

1. Avoidance of assumptions related to the effects of largely applied interventions;
2. Inclusion of lesser known and non-traditional dimensions and interventions of well-being; and
3. Generalizability to decision making across all the sectors.

A tool of synthetic mapping is used for describing the current state and breadth of the evidence base with the key intention of characterizing and visually representing the casual associations supported by the evidence base.

Systematic mapping, which is also known as evidence gap mapping is a very useful method to synthesize and characterize thematically, or map the collections of evidence especially in order to provide a base for policy decisions.

This method allows the identification of weak and lacking evidence on the major associations that might enable policy-making decisions based on the current knowledge so as to direct the resources in an effective manner.

The outputs of mapping comprise structured matrices that represent the distribution as well as the frequency of the studies, and coding for designing the study and mechanism across a broad variety of interventions and the outcome variables.

Apart from characterizing the current evidence base, the theory of change conceptual modeling methods is applied to define the categories for the organization of data for example, long-term, and intermediate outcomes, and also to interpret the direct as well as indirect casual pathways between interventions of conservation and outcomes of human well-being.

7.4.1. Aims and Objectives of the Systematic Map

The mapping method or approach will scope and recognize the studies that measure and document the effects of nature conservation intervention on the well-being of humans at the local and regional scales both.

The key focus will be on the direct impact on discrete populations, targeted stakeholders and beneficiaries, along with an emphasis on individuals, households, communities in all the developing nations as all these populations are highly dependent on the nature and they are very much sensitive to the positive as well as negative impacts of conservation.

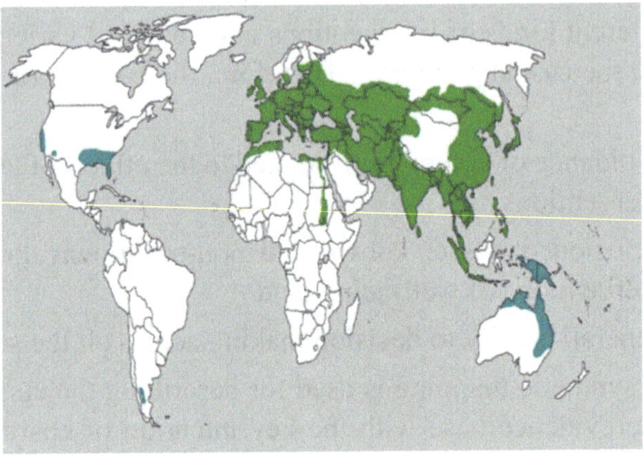

Figure 7.2. A systematic representation of the distribution of wild boar on a map.

Source: Image by Wikipedia.

The key research question arises is, what are the impacts of nature conservation interventions on variety of domains of human well-being in the developing nations? The systematic map will construct an index of literature and plot the evidence demonstrating the breadth of the casual associations between positive as well as negative effects of the nature conservation and well-being of humans.

The main objective of the systematic map is to evaluate and characterize the existing state and the distribution of the current evidence base. The map is essentially focused on the evidence that are related to the effects of nature conservation that comprise interventions mainly intended to attain the protection as well as management of the natural ecosystems in-situ.

This, in general, excludes ex-situ interventions like seed banks captive breeding and the environmental regulatory measures like control of waste management, air pollution, energy production and water treatment.

Well-being of human is a multi-dimensional notion, including both tangible as well as intangible effects and broadly defined as the 'satisfaction of human requirements to attain a state of well-being that is happy, healthy, and prosperous, both mentally and physically.'

The evidence base will be further explored and categorized using a targeted extraction strategy for data to answer some of the secondary research questions such as:

1. What is the existing state and the distribution of evidence?
2. What kinds of impacts from conservation interventions on the well-being of humans are measured?
3. What kinds of ecosystem services are linked with the impacts of conservation interventions on the well-being of humans?
4. Which of the populations are highly affected by the conservation of wildlife?
5. How does the evidence base line up with the major priorities and the investments of executing agencies?

All these consists of recognizing the variety of conservation interventions that are associated with distinct domains of human well-being together with the duration of interventions, geographic locations and the kinds of conservation aims.

7.5. HOW SAVING WILDLIFE BENEFITS HUMANS

It is not a secret that earth has lost a great number of species in the past four decades. All these species have been disappeared in the past because of deforestation, overpopulation, climate change, consumer culture, animal exploitation and various other harming sources that were inflicted by humans, mankind. It has been said that around 10000 species extinct every year, halving the wildlife population of the whole world in just the past 4 decades. One of the major rules of terrestrial biology is 'no species is forever,' according to Jeffrey Kluger. Although, such rapid loss of species every year is estimated by the experts to be 1000 to 10000 times higher than the natural extinction rate. As highly accepted theories have agreed and as the Science papers show- people are now in the middle of the sixth major extinction, the unsettlingly-named Anthropocene, or the age of the humans.

When the public and politicians started insisting on the protection of environment in the year 1960, the laws of conservation began passing in the year 1980, beginning with the Alaska National Interest Lands Act, which set aside almost 101 million acres of Alaska to be preserved as a monument, national park or wildlife refuge in order to maintain one of the United States' last "wild" areas. With high involvement of the public in wildlife conservation, restoration, and preservation of the endangered wildlife species became possible eventually. As WWF puts it perfectly, wildlife conservation is a key source of inspiration.

The natural assets of the earth are made up of land, water, plants, animals, the atmosphere and obviously the human beings. WWF goes on to assert that, biodiversity supports the health of the planet and has a direct impact on the lives of people. In simpler terms, damage to biodiversity is forcing millions of people to face a future where the food supplies are vulnerable and where the freshwater is short or irregular supply. If biodiversity affects the lives of people directly in big ways, then it is safe to say that efforts of conservation do not only benefit the environment but to human beings also.

7.5.1. Food Security

One key benefit that evolves out from the efforts of wildlife conservation is that it assures the security of food. Protection of forests from deforestation and rebuilding forest habitats to protect biodiversity aids in the carbon-sequestering process, offers new economic opportunities, and guards against erosion.

Figure 7.3. Wildlife conservation ensures balance in the ecosystem and food security.

Source: Image by Pixabay.

In addition to that, conservation of wildlife promotes agricultural biodiversity that plays a key role in building a robust, secure, and thriving system of food.

When the agricultural biodiversity is exploited, the land is clear for agricultural practices and huge habitat loss takes place along with undocumented loss of species and huge solid erosion. Several researches show that this process has a negative impact on the health, nutrition, and dietary diversity of certain groups of the society.

7.5.2. Public Health

Another key benefit that comes from the conservation of wildlife is that all these initiatives secure health of human. CT reports that much more than 50% of the modern medicines and more than 90% of the conventional medicines come from wild plants and animals.

These conventional medicines hence, demonstrate an essential pharmacopeia and the body of medical knowledge that cannot be replaced by the synthetic alternatives in an easy way.

In addition, a world that encourages biodiversity, as well as healthy ecosystems offers critical buffers between humans and diseases.

A great number of studies have associated reduction in diversity amongst the mammal species and total reduction in biodiversity to an increase in the transmission of animal-borne diseases to human beings. Biodiversity and healthy ecosystems regulate the change in climate and mitigate air and water pollution in an effective manner.

7.5.3. Creating Opportunity

Possibly, the most convincing benefit that arises from the conservation of wildlife is that it provides people with an opportunity, whether the opportunity is social, economic, and cultural.

The industries working for unsustainable resource extraction- such as the bush meat industry, the charcoal industry, the poaching industry and the clear-cut logging industry-extend the gap between the rich and the poor. It has also been associated with political strife and civil war.

Additionally, an increase in healthy ecosystems and biodiversity by conservation eventually enhances the agricultural productivity, hence allowing agricultural farms to become profitable. A healthy ecosystem in which a great number of unique species live in all around the world that aids the local economy and it also invites a completely new fusion of investment.

The projects related to wildlife conservation bring the communities all around the world together, which is usually called as a team effort. Starting from rangers to the administrators, the initiatives of wildlife conservation invite the opportunity for a novel economy based on conservation.

An unconscious, unsustainable, and self-interested relationship with the environment has effectively led people into an exceedingly destructible world. If people will not take active action and go about changing their ways, then people are at risk of losing very crucial and irreplaceable biodiversity and ecosystems.

7.6. SAVING THE SPOON-BILLED SANDPIPER IN BANGLADESH

The first specimen of the Critically Endangered Spoon-billed Sandpiper (*Eurynorhynchus pygmeus)* was collected in mid-19th century from Bangladesh. The country holds the record for the highest single count of the bird all around the world- 202 birds in the year 1989 from Moulevir Char (Bakewell and Howes, 1989).

Figure 7.4. Spoon build sandpiper.

Source: Image by Wikimedia Commons.

Since then the birds has undergone an extreme reduction in their population. As per Birdlife International, this charismatic species is listed as "critically endangered" as these are present in very small number. Also, it is undergoing a rapid decrease in its population.

This is mainly because of several crucial factors such as habitat loss in its breeding, wintering, and passage grounds; all these are compounded by hunting, disturbance, and the effects of change in climate. Juvenile recruitment and fledging success are quite low which leads to fears that the population is aging in a very rapid way. Therefore, now urgent action is needed to be taken in order to prevent the extinction of this species.

This species has a limited breeding range naturally in north-eastern Russia (Bird Life International, 2001). It drifts down the western Pacific coast by Japan, Russia, South Korea, North Korea, mainland China, Taiwan, and Hong Kong (China) and Vietnam, to its key wintering grounds in Myanmar and Bangladesh.

Considering its ugly global status, a group of young ornithologists in Bangladesh took up the challenge of saving this highly threatened species in the year of 2009. It was the time when the Bangladesh Spoon-billed Sandpiper Conservation Project (BSCP) with a small grant from The Explorers Club started.

After that, regular Spoon-billed Sandpiper surveys were conducted all over the winter season in possible historical sites, with major support from the Spoon-billed Sandpiper Task Force and other international organizations.

In March 2010, a team found a minimum of 25 Spoon-billed Sandpipers on Sonadia Island off Cox's Bazar and 23 on Domar Char. These were the highest counts in Bangladesh since the phase of the 1990s, and further monitoring has also confirmed that these two sites support similar numbers of birds in a regular fashion.

The surveys basically confirmed that Bangladesh is still a very significant wintering ground for Spoon-billed Sandpiper. It has also been recognized that Sonadia Island as the major wintering site, while Domar Char as a stopover (passage) site in Bangladesh.

At the time when researchers used to work on Sonadia Island, they found certain evidence of shorebird hunting. The research specifically pointed towards the eight hunters who claimed that they basically target larger shorebirds like Whimbrel or Curlew, they had captured a total of 22 Spoon-billed Sandpipers in winter 2009–2010.

From September in the year 2010, a series of surveys were conducted in order to recognize hunters and some alternate options for them Sonadia Island. The BSCP signed the agreements related to conservation with a total of 25 active shorebird hunters on the island during October to December in the year 2011. Essential resources were offered for alternate livelihood options such as fishing nets, boats, livestock, and equipment for the cultivation of watermelon, and tailor and grocery shops.

Village Conservation Groups (VCGs) have been established in total of five villages in and around Sonadia Island. They observed the new activities related to livelihood and the former hunters repay a small portion of the income generated to their VCG.

The VCGs use this money for hunting mitigation awareness activities related to shorebird conservation within their villages. The complete process is being observed and guided by the BSCP. The hunters who have taken up some alternate livelihoods agreed for the protection of birds from any kind of threat, such as hunting and habitat destruction, and also to assist the conservation of nature in their villages.

Few of them took the accountability for securing and observing the activities regarding bird hunting at some of the nearby shorebird sites and all the former hunters visit these sites a day a week in rotation. Future actions will comprise a further awareness campaign targeted at the five villages at Sonadia Island, sustained monitoring of the activities of former hunters, and surveys along with villages to examine the scale of shorebird hunting.

Regular observation of shorebirds will be done on Sonadia Island in order to locate wintering, passage, and over-summering Spoon-billed Sandpipers, and new areas of the Bangladesh coast will be searched in the Sundarbans, Padma-Megha delta and Chittagong.

According to Mohammad Foysal, Senior Research Officer of BSCP, "We are confident that bird hunting has stopped on Sonadia Island after our conservation agreement scheme and people will be further motivated to save this species by our forthcoming conservation awareness campaign."

Ultimately, the situation has improved further for Spoon-billed Sandpiper in Bangladesh from hunting pressures. However, a recently proposed deep sea port on Sonadia Island could be an extreme threat to this species in the future.

7.7. CONCLUSION

Sustainable conservation of wildlife by implementing certain practices does not take by itself. Actually, various factors can work in this direction. The need of conserving wildlife plants and animals are quite different from those of conserving complete biodiversity. The conservators may try to prioritize the expectations of people living around the world. Competition for effective resources between people may cause overexploitation of resources.

Implementation of conservatory practices for the conservation of wildlife is all about using strategies and policies that represent different ways as well as means of attaining set goals for wildlife conservation. Policies usually defer to institutional setups that allow the governance of wildlife conservation.

Policies may not be very specific and they may apply across in every area of a specific country while the implementation of strategies are much action oriented and are associated to a region or destination. Both policies and strategies define a desired state of the future along with detailed analysis of important steps to attain that projected vision.

However, the government is taking essential measures. It is necessary to have a constant interrelationship between different strategies and policies. In order to attain the goal of wildlife conservation, several tools like policies can be used along with certain partnerships that allow effective activities for the conservation processes and training or awareness programs.

REFERENCES

1. Bottrill, M., Cheng, S., Garside, R., Wongbusarakum, S., Roe, D., Holland, M., Edmond, J., & Turner, W., (2014). What are the impacts of nature conservation interventions on human well-being: A systematic map protocol? *Environmental Evidence*, [online] *3*(1), 16. Available at: https://environmentalevidencejournal.biomedcentral.com/articles/10.1186/2047-2382-3-16 (accessed on 21 August 2020).
2. Chowdhury, S., (2016). *Saving the Spoon-Billed Sandpiper in Bangladesh*. [online] Conservation India. Available at: https://www.conservationindia.org/case-studies/saving-the-spoon-billed-sandpiper-in-bangladesh (accessed on 21 August 2020).
3. Essays, Research Papers and Articles on Environmental Pollution, (2020). *Essay on Wildlife Conservation (671 Words)*. [online] Available at: http://www.environmentalpollution.in/essay/essay-on-wildlife-conservation-671-words/158 (accessed on 21 August 2020).
4. Hill, A., (2020). *Wildlife Conservation Projects Do More Harm Than Good, Says Expert*. [online] the Guardian. Available at: https://www.theguardian.com/environment/2010/jul/29/wildlife-conservation-projects-more-harm (accessed on 21 August 2020).
5. Qutab, M., (2020). *How Saving Wildlife Benefits Humans-In Ways We Really Need*. [online] One Green Planet. Available at: https://www.onegreenplanet.org/environment/how-wildlife-conservation-benefits-people/ (accessed on 21 August 2020).
6. Springer, J., (2009). Addressing the social impacts of conservation: Lessons from experience and future directions. *Conservation and Society*, [online] *7*(1), 26. Available at: http://www.conservationandsociety.org/article.asp?issn=0972-4923;year=2009;volume=7;issue=1;spage=26;epage=29;aulast=Springer (accessed on 21 August 2020).

8

Future of Wildlife Conservation

CONTENTS

Wildlife Conservation is an important part of wildlife management and a very important matter to look into. The future of the ecology and the biodiversity of the world depends on the measures that may be taken to improve the wildlife conservation measures. This chapter takes the readers through the future possibilities and results of wildlife conservation. It dwells upon the various aspects of wildlife conservation and its effect in the future.

8.1. INTRODUCTION

In tropical forests, human extractive activities including hunting but not only restricted to it are troublesome procedures and can trigger various, yet not totally understood, mechanisms (compensatory or predation rate changes) or impacts (a trophic cascade or keystone impacts) which will, in turn, modify, in a significant way, the Overall composition function and structure of the ecosystem.

However, each organism contributes to the processes of the ecosystem, the magnitude of individual species and nature contributions vary considerably. Most ecosystem processes are driven by the joined exercises of numerous species.

Plant regeneration (seed dispersers, seed predators and loss of pollinators), plant diversity (increased pests, change in herbivory patterns), food webs (loss of top predators or of their prey), are amongst the potential to impact the ecosystem more broadly and not only the targeted species.

Figure 8.1. Predators have been diminishing in numbers which has disturbed the food web.

Source: Image by Pixabay.

Various species, who have similar tropical status or life history and perform similar roles in ecosystem processes constitute what has been termed as functional groups. Species within these groups, for example, large predators, nitrogen-fixing microbes, grazing mammals or perennial grasses are functionally similar in spite of their uniqueness in life history, genes, and various other attributes.

As several species contribute in similar ways, it is along these lines usually hard to determine relative contributions of a given species to the ecosystem processes. However, some functional groups or species matter more than others.

This turns out to be particularly clear on account of 'keystone species' which are even alluded to as 'ecosystem engineers' or organisms with high 'community importance values.' All these terms allude to species whose loss disproportionately affects the community when contrasted with the loss of Various other species.

Conventional wisdom thusly predicts that as hunters are more inclined towards large animals those are usually the keystone species; the extirpation or reduction of these animals will bring about dramatic changes to the ecosystems. A portion of these anticipated changes has been exactly shown while others presently have so far proved to be inexact or have yet to be demonstrated.

Wildlife resources comprise an indispensable connection in the endurance of the human species and have been a subject of a lot of interest, research, and fascination everywhere throughout the world. Today, the effective conservation of wild animals is of great significance since a large number of species of wild fauna have become endangered, and wildlife habitats are under severe pressure.

Since all of us rely upon animals and plants for All significant components of our welfare, it is more than a matter of convenience that they keep on existing; it is a matter of life and death. Being living units of the ecosystem, animals, and plants add to human welfare by giving:

- The material advantage to human life;
- Information about genetic resources and their conservation; and
- Remarkable contributions to the enjoyment of life (e.g., recreation).

Human society relies upon genetic resources for effectively all of its food, almost 50% of its medicines; a lot of its apparel; in certain areas, the entirety of its building materials and fuel; and part of its spiritual and mental welfare.

Figure 8.2. Humans rely upon wildlife for a lot of materials of their consumption.

Source: Image by Wikipedia.

Considering the manner in which we are dashing ahead, unmindful of what legacy we intend to leave for people in the future, the future doesn't appear to be too bright. Analysts have anticipated that by the year 2020, the human populace will have expanded by more than half, and the tropical forests and the arable fertile land will reduce to less than 50% of what they are today.

We need to care about genetic resources although they are treated as inexhaustible mineral resources. It is here that the idea of the preservation and management of wildlife comes into play since anything that isn't human or undomesticated is 'wildlife.'

The absence or presence of plants or animals in an area is controlled by historical and ecological factors. Plants and animals are considered the living indicators of the quality of their environment. their extents mark the regions where environmental conditions are equivalent or comparable.

To explicate the scope of species appropriately, it is important to know, in detail, the conditions required for the species to live and flourish. The science of zoogeography has both historical and ecological angles. On this basis, the world can be separated into six zoogeographical zones:

- Nearctic — North America and Greenland
- Palearctic — Eurasia, without India
- Ethiopian — Africa, south of the Sahara
- Oriental — India and Indochina
- Australian — Australia and New Zealand
- Neotropical — South and Central America, and the Antilles

8.2. CURRENT TRENDS IN WILDLIFE CONSERVATION

In the current situation protection of biodiversity is a great challenge. In the modern changing world, it requires shared Global endeavors for success. There is a great need for various planning strategies for conservation, protection, and limiting the loss of natural resources. Sign of preservation targets or endpoints and cautioning with site-specific significantly upgrade the achievement pace of protection.

8.2.1. Objectives

- To assess the present threats to wildlife.
- What are the ongoing endeavors that have been accomplished for the preservation of wildlife?
- Feature the potential endeavors ought to be accomplished for saving wildlife in the next future.

8.2.2. Main Causes of Extinction and Endangerment

1. Loss of Habitat and fragmentation.
2. Wildlife Trades by National and International concern.

- Trade of coral reef;

- Trade of pets;
- Trade of body parts;
- Trade for biomedical research; and
- Trade of bushmeat.

3. Other causes.

- Climate change or global warming;
- Pollution;
- Introduction of invasive species; and
- Rancher or farmer shooting.

Presently, in Asia, wildlife species being evacuated or probably because of human mistreatment at an alarming rate. All the more as of late ecological and conservation researches published in principle journals are one-sided in the direction of study sites based on intemperate and affluent nations.

Southeast Asia is perceived as one of the most sizzling of the world's hotspots of endangered biodiversity. The primary risk for Asian wildlife is hunting and critical habitat loss.

In this way, for management, protection, and conservation of wildlife, longitudinal research is essential particularly on account of the unlawful and lawful trade, and to know about the fundamental population dynamics, identified with long live species.

8.2.2.1. Loss of Habitat and Fragmentation

Habitat loss is the greatest threat to biodiversity. During the most recent two centuries, the human populace increments quickly so as to meet their necessities like farming, land, and urbanization a significant measure of natural vegetation was debased which prompts deforestation around the globe.

For biodiversity conservation Forests assume crucial and driving job that is perceived by the global community and prioritized forest biodiversity conservation and preservation through various agreements and projects, for example, Aichi's Targets and Millennium Development Goal.

In the year 2010, the five parties of the Convention on Biological Diversity (CBD) sorted out a deliberate arrangement for 2011–2020. This plan or arrangement is known as Aichi's Target and has five vital objectives.

It is stepping forward in 2010 by the first objective "accomplishing noteworthy decrease of the present pace of biodiversity lose." The handling of twenty targets at the same time is very tuff and has great burden for certain nations.

Although progress is low yet in October 2014, the twelfth conference of the parties was held and consented to improve progress to accomplishing the 2020 objective. The forest area that is intended for preservation of biodiversity we can evaluate that present zone as a region of forest that is secured, and forests that are as of now present in nation and biomes.

According to the recent studies Earth's 30% of the land area is covered by forests having 80% terrestrial biomass and almost more than half of the terrestrial animals and plants biodiversity of the world inhabiting there.

However, internationally perceiving the importance of forests, and admitting that the area of forest has been contracting consistently because of fragmentation while the land of agriculture is growing in 70% of nations.

During the 1990s it is assessed that every year 16 million ha of deforestation is decreasing world widely and between the year 2000 and 2010 is 13 million ha yearly however on the other hand recently 0.08% rate of loss is accounted for the years 2010 to 2015 is at a tranquil issue of to be concerned in light of the fact that there is an enormous loss in the zone having high biological value.

Since a lot of a number of known terrestrial species live in forests, and currently, around 9% of trees diversity alone is as of now in danger of extinction, so reversing or slowing the deforestation is the main objective of Aichi Biodiversity in addition to Millennium Development goal is also very significant.

As the world's community knows about the importance of biodiversity and woods spread, along these lines there is a great improvement toward accomplishing biodiversity protection and limiting the loss of forest goals but it appears to be a big test.

Presently, one of the essential methodologies for the century is assigning protected areas (PAs); Although, it is discovered that a huge number of significant biodiversity regions are not satisfactorily covered as secured region rank. Among protectionists amplifying forest region is a major concern.

At Global-scale Forest Resources Assessment (FRA) offers opportunities to gather and analyze information. Because of this pattern, numerous issues that control wildlife conservation are featured at large scale and propose numerous probabilities for protection and enhancement of biodiversity.

At the point when analyzed according to different concerns (e.g., socioeconomic data), that offer approaching into the perspectives that impact the conservation of forests at a huge scale and recommend promises for development together in the conservation of forest and in following the position of forests which is very important for conservation.

8.2.2.2. Global Forest Area

In the year 2015, it revealed that the total worldwide forest area is around 4000 million ha, while in the year 1990, the total area was 4128 million ha, so it is reported that there is a decline of about 3%. The fundamental driver of this reduction is the decline of the tropical domain that is around 200 million ha, and there is 65 million ha partly increase in the temperate domain.

In 2015 it is proclaimed that in the tropical domain there is the highest percentage of forest area that is 44% of aggregate, and temperate domain followed as 26%. Region-wise, North America has 16% forest area, South America has 21% forest area, and Europe has the highest forest area 25%.

8.2.3. Role of Planted Forests

A significant role is played by planted forests in biodiversity conservation and it also reduces the pressure on natural forests. During 1990–2015 it is evaluated universally that afforestation is increased and deforestation is decreased.

Both Food and Agriculture Organization Corporate Statistical Database (FAOSTAT) and FRAs data show that if globally planted forests increments as 2.4% per annum from the year 2010 to the year 2050, natural forests for fiber as well as timber might be restored.

The current 80% of the forests are added to the production. That isn't different from the 2.05% per annum increment revealed by FRA for 1990–2015. There is an extraordinary need for awareness in public in this time for the conservation of wildlife and forest. Consequently, Tourism is the most ideal route particularly for university, college, and school students.

For biodiversity conservation forests designed as a protected area

For lessening the negative effect of human maltreatment on the natural environment many secured territories on the planet are set up for keeping up conservation of wildlife and maintaining balance.

As indicated by IUCN in FRA "region particularly devoted to the maintenance and protection of biological diversity, and of natural and associated cultural resources and managed through legal and other effective means is known as protected area" just as "Forest area planned fundamentally for the conservation of biological diversity that incorporates, however, isn't constrained to zones assigned for biodiversity conservation within the protected regions is known as the area for conservation of biodiversity."

In 2010, a maximum of 86% of forests intended for biodiversity conservation 234 nations announced as officially protected forests. In the year 1990, globally more than 80% of the forest area is reported to be protected and the zone for the protection of biodiversity is 88%.

The various countryside does not give reports in all years but has large PAs including Mexico, Australia, Venezuela, Congo, Peru, and Bolivia that is the reason it is difficult to compare data between years. In the year 2015, out of the total 651 million ha of protected forests, only 507 million ha area of forest is reported as protected.

Over 25-years there is a territory that is secured and expanded in every single climatic sphere, the majority of the increase took place before the year 2010. About 0% was in 1990 to 15% of backwoods secured region in 2015.

There was 0% of the forest protected area in the year 1990 and it increased to 15% of the forest protected area in the year 2015. List of top 15 nations which have disclosed the largest protected area in the year 2015 are the United States, China, the Democratic Republic of Congo, Australia, Russian Federation, Botswana, Pluractional state of Thailand, Zambia, Brazil, Indonesia, Venezuela, Canada, Peru, India, and Bolivia. Various nations are not included because they did not give reports consistently.

In 2015, in all nations out of the total 524 million ha area, 427 million ha area was designated as an area for biodiversity conservation. Conservationists have extraordinary worry for maximizing forest area however the significant issue is climate change.

Political inclusion guaranteed that there is zero deforestation and international financing constantly demanding that decrease the deforestation and increment the afforestation. In recent years the biggest recognized problem is drought, which increases the mortality rate in forests of Western North America.

In shrub-land and grasslands, habitats are constantly decreasing due to disturbance. To handle this decrease Government agencies, utilize anthropogenic disturbance like logging or mowing to emulate nature once.

Another progression taken for land management is the management of database by Massachusetts that will handle all land management exercises by state organizations. This database is profoundly huge. Permit coordination among NGOs and various organizations and give simple access to data for management efforts.

8.2.4. National and International Wildlife Trade

Wildlife trades are a great threat for the conservation of wildlife. Convention on International Trade in endangered species (CITES) depending on trade bans and controls.

- **Trade of Coral Reef:** Ornamental coral reef wildlife trades incorporate the aquarium, curios, and jewelry give multi-million dollars to the industry. Vulnerable coral reef species are threatened because of not great collection practices and absence of effective management and guidelines. The primary international mechanism that is an understanding among 175 nations that control the coral reef trade is CITES.

8.2.5. Trade of Pets

The welfare of animals is a global concern and keeping wildlife animals as pets is an illicit trade for conservation. Numerous national and international Non-governmental organizations (NGOs) are actively working to diminish this unlawful trade.

In the domestic market of Peru, it is perceived that 28 NGOs are working. In an interview, it is seen that only 5 NGOs that have solid point of view toward wildlife conservation have high endeavors to diminish this unlawful trade in two regions of Peru. In general, interviews indicate that there is a reckless conduct of government toward legislation.

Figure 8.3. The animals are traded as pets or for places such as zoos.

Source: Image by Wikimedia Commons.

8.2.6. Trade of Body parts

All around the globe, wildlife species are hunted for specific parts of animal bodies that are utilized in fetish practices and medication. Animal parts are likewise utilized for cultural and traditional practices in Africa.

Avian mortality identified with human persecution for trade in the Cross-River region of southeast Nigeria is seen through personal interviews from 18 villages conducted in 2015. Outcomes indicate that 27 bird species from 13 families utilized as regional trade out of which top 5 are globally perceived as threatened species.

Figure 8.4. Pangolin trade.

Source: Image by Wikipedia.

- **Trade for Biomedical Research:** There are numerous laws, conventions, policies, regulations, organizations, and treaties for the utilization of animal species for biomedical research yet additionally there are further prerequisites for non-human primates. These prerequisites incorporate political activities, moral, and ethical concerns, medical needs, cultural, and social perspective, and species differences.

8.2.7. Bushmeat Trade

In tropical regions utilization of wildlife is huge, local communities consume around 5 million tons of bushmeat. For conservation of biodiversity human conduct has incredible significance. For conservation project compliance about conservation rules (e.g., no chasing, shooting, and so forth) is exceptionally important.

Since conservation is a very delicate issue and victims are voiceless so "detection and understanding violation of conservation rules" are vital. Because of this, it is clearly a need to understand issues for better complaints and effective implementation.

8.3. INTERNAL AND EXTERNAL INTEGRATED CONSERVATION

Aquariums and zoos started to understand their potential as a positive and powerful power for conservation of wildlife as right on time as the beginning of the twentieth century, and by the 1960s progressively included conservation as a significant part of their overall mission.

All through the world, there are aquariums and zoos, especially those in WAZA, that assume an amazing job in the conservation of biodiversity and they endeavor to expand their contribution to global conservation in different ways.

The goal now of the world zoo and aquarium network is that every one of its individuals is directly connected with conservation programs in the wild and seems to be included. Without integrating conservation into its organizational culture, no individual aquarium or zoo can contribute to conservation in a meaningful way.

Integrated conservation must be its explicit and clear aim. Integrated conservation is accomplished most adequately when all the exercises of a zoo or an aquarium are connected to each other adroitly, and are strategically coordinated both internally and externally.

Their primary aim being the maintenance of healthy ecosystems and conservation of threatened species. The procedures of collaboration, communication, and coordination should become easy and routine. In this manner, integrated conservation incorporates a lot of internal procedures by which a zoo attempts to deal with all of its exercises and relationships in support of well-defined and specific conservation programs.

Integrated conservation can likewise fill in as a banner under which conservation programs can be conveyed to zoo visitors, the media, the general public and the supporters. Integrated conservation activities will vary in various parts of the world as a result of the everyday reality of life and social and cultural factors.

Zoos situated within the high biodiversity regions regularly contribute quite a bit of their time, financial resources and energy in giving treatment and holding territories for individuals, formerly free-ranging, wild animals.

These incorporate those animals which collide with individuals in urban towns and cities, those which have been intercepted in unlawful trade, wild animals are taken as pets which have become a burden for their owners, or those who become victims of natural disasters, for example, floods, earthquake or fire, or even those which are lost, have wandered or have been taken.

These zoos are regularly vigorously associated with welfare issues and this can influence the manner in which the organization takes part in, or even explicates, integrated conservation.

Zoos and aquariums around the globe can attempt field conservation both in their own nation and abroad. Numerous zoos in North America and Europe do quite a bit of their conservation work abroad, especially in high biodiversity nations, just as working within their own regions, though the aquariums and zoos of Australia, a territory that incorporates a portion of the hotspots for endemics, collectively directs more conservation assets to within region projects than to outside region projects.

Numerous aquariums and zoos in high biodiversity nations, for example, in South and East Asia, Africa, and Central and South America, are as yet attempting to set up their own roles in integrated conservation.

These foundations regularly have significantly larger visitor numbers than zoos somewhere else and these can use quite a bit of their staff's energy and work. Such zoos are, although, ideally situated to educate a huge number of individuals about their nation's conservation potential and problems.

Subsequently, the idea of integrated conservation may mean various things in various places. There is constantly potential competition for assets between a zoo's income creating exercises and its conservation needs and desires.

An aquarium or a zoo which is financially successful will consistently have the option to allot or collect more money for conservation than a zoo which is battling to meet its responsibilities to its own staff, animals, and guests.

Although any aquarium or zoo, short of surplus money, can find significant approaches to contribute to conservation. At last, we would anticipate that, as opposed to creating competition, the conservation roles of aquariums and zoos will turn out to be so very much coordinated with institutional achievement that good conservation will actually enhance the institutional budget.

Integrated conservation falls into two different however related sets of exercises, external, and internal. External activities are those which an institution conducts away from its own grounds and internal activities are those which an institution conducts itself and acts in regard to its day-to-day visitor-related actions.

8.3.1. Internal Integrated Conservation

Most aquariums and zoos around the globe as of now perform numerous activities that would be depicted as parts of internal integrated conservation. A portion of these are recorded underneath.

- All aquariums and zoos care for animals and exhibit them in enclosed areas. Sometimes they build groups of enclosures, regularly connected by a biological or conservation-related topic that depends on ecosystem, habitat or geographical location. At times, exhibits contain several different species, both plants and animals.
- Aquariums and zoos fill in as recreational facilities for social groups, families, and people everywhere throughout the world. In numerous parts of the world, they are one of the significant sources of affordable and safe outdoor entertainment.

- Through graphics and various other methods of interpretation and engagement, for example, animal feeds, natural behavior animal shows, and keeper talks, they can Explain the behavior of animals and biology within the enclosures, including social conduct, population ecology and reproduction. A few zoos additionally explain about the wild habitats where the animals are found, the dangers confronting the species in the wild and what zoos are doing to support conservation.
- Many zoos have education departments for both informal and formal educational processes, in spite of the fact that in certain parts of the world, education is undertaken by staff that have various other obligations or even by local NGOs.
- In order to draw in visitors, zoos showcase themselves to the overall population utilizing numerous strategies, including word of mouth, advertising, and public relations. In certain nations, although advertising is not really vital and even restricting numbers at certain times may have to be considered.

In future, by adopting a strategy of integrated conservation, aquariums, and zoos will also:

- Adopt a really maintainable way to deal with their own construction and maintenance processes by, at every possible opportunity, building with recycled materials or sustainably sourced with low embodied energy, limiting energy consumption by utilizing passive heating systems and insulation. Generating their own energy by utilizing techniques, for example, wind power and solar gain, and explaining all these 'green' initiatives to their guests.
- Make explicit connections between the entirety of their significant exhibits and field conservation projects so that guests find out about the conservation status of the animals they are taking a gander at.
- Engage the broader community and visitors in discussion about the more extensive issues threatening species in the wild and attempt to move them and in this manner secure their help.
- Attempt to connect catering facilities and shops with conservation programs-for instance, by selling crafts from the zone of a conservation programme and utilizing the returns to profit local individuals from that region;

- Inform visitors about the conservation work of the aquarium or zoo in addition to that of different aquariums and zoos, government agencies and conservation organizations.
- Draw conservation issues to the consideration of a wider crowd through promotional work, for example, the internet, advertising, and public relations. The internet should not be underrated as an instrument for building consensus and expanding mindfulness on conservation issues.

8.3.2. External Integrated Conservation

The WZACS accentuates that the complex, modern world of conservation has numerous players and numerous agendas. No single association, be it aquarium, development organization, zoo or conservation charity should act alone.

Preservation exercises ought to be shared, with all the partners moving in the direction of a similar end, and maintaining a strategic distance from exploitation or competition.

Conservation-active aquariums and zoos must collaborate proactively with human development organizations, national, and international conservation agencies, government divisions and local communities, to guarantee long haul sustainable solutions.

In the past, many conservation activities have failed to take enough account of the wider agenda specifically in human development and this is still of significant concern.

Aquariums and zoos are popular visitor attractions and thus have unique opportunities to introduce their visitors to a wider world and to explain the issues of international conservation, unlike many conservation organizations which are not highly visible to the general public.

They can enormously improve guest's awareness of conservation matters, both solutions and problems, by incorporating their own work with that of other conservation bodies; by indicating proof of that integration, they become the 'shop window.'

They can likewise act as physical foci for integrated networks of conservation and development organizations, giving central assets, for example, training, and meeting facilities.

8.4. CONSERVATION AREAS OF CONCERN

It is difficult to list all of the conservation regions that conservationists, special interest charities, researchers, and governments. The list given below incorporates some of the most pressing, urgent or significant to conservation today.

8.4.1. Heritage Crime

If we go back just a few decades, the biggest threat to heritage conservation was indifference and absence of concern for conserving the past in striving for economic growth.

Today, we cherish our heritage undeniably more than we used to, credit goes to the travel industry and public perception of our built heritage. Be that as it may, in certain regions of the world, political instability has led to a growth in what is classified as 'heritage crime.' There are various perspectives to this:

- **Iconoclasm**: the purposeful targeting of artifacts and monuments for destruction as a symbol of a people's cultural identity based on the past, or the destruction of religious minority sites. The circumstance in the Middle East has seen huge harm to the significant ancient Roman and Greek sites in Syria on account of ISIS. So also, the Taliban annihilated Buddhist sculptures in Afghanistan.
- **Art and Antiquity Theft**: the toppling of President Hosni Mubarak in Egypt and the fall of Saddam Hussein's regime in Iraq led to the huge theft of priceless and irreplaceable artifacts (33) that had become symbols of every nation's cultural heritage. Sometimes stolen to order by rich donors, a lot more are opportunistic criminals wanting to sell it through the black market, taking advantage of ransacked museums and rioting.
- **Damage to Culturally or Archaeologically Sensitive Land**: It can likewise incorporate illicit excavation for archeological treasures, digging or mining for assets for economic benefit, which is some of the time illegal if the land is exceptionally secured for heritage conservation or unlawful building of residential, commercial or industrial properties without consent leading to harm of a site heritage. Historic England, an NGO responsible for monitoring and maintaining historic sites in England reported

that around 18% of the secured heritage conservation sites and regions endured heritage crime in the year 2017.

Sometimes it is not possible to repair when damage occurs. In the instances of the lootings in Cairo and Baghdad, there are as yet numerous antiques missing that global authorities are attempting to find.

8.4.2. Expansion of Boreal Forest in the Arctic

In most cases, the growth of forests without the intervention of ecologists and conservators will, by, and large, be viewed as something to be thankful for. All things considered, tree cover can only be useful for biodiversity and as a carbon net sink.

However, in an ecologically delicate region as the arctic circle, the spread of boreal forests on the landmasses of northern Russia and Scandinavian more temperate areas is presently of profound concern. Environmental change has led to warming in these regions.

Arctic boreal tree spread is presently spreading at a terrifying rate and undermining the presence of an enormous number of native species that were at that point in a sensitive conservation state.

It is likewise accepted that the spread of tree canopy has triggered the warming previously being initiated by carbon emanations. A procedure that has been seen to take hundreds of years during warming stages in the past is currently taking decades.

8.4.3. Energy Conservation

As the populace grows, we need more vitality. Fossil fuel supplies are diminishing, and many have accepted that we have already reached and passed peak oil. Steps are as of now being taken to decrease our reliance on these fuel types, yet in the western world, funds, and grants for renewable energy are confronting government reductions even with industry campaigning.

However, there are various steps. our gadgets require far less power than they did 10 years ago. Our premises and homes are turning out to be more energy-efficient, however, the amount of energy we use keeps on rising, even after traditional bulbs replaced by LEDs, conventional battery power replaced by power cells, and continuous evolution of renewable energy.

The Energy Information Administration predicts a worldwide ascent in energy utilization of 56% between now and 2040.

8.4.4. Food and Water Security

The ecological problems currently faced by us, for example, flooding, ice cap melting, climate change, ocean acidification, drought, environmental change, and numerous others have knock on impacts for our food and water security.

Drought and flooding both have the ability to harm our food supply, and in various cases securing the wellsprings of both are conservation issues. In the case of water, there should be a lucid arrangement for diminishing the amount of wastage and motivating households and businesses to lessen the amount they use, and not only during drought periods.

Various countries and few states in the United States have encouraged the utilization of water buttresses to gather rainwater to use to flush toilets and for plants, for instance. Food security is reliant on water security, yet it tends to be dependent upon meteorological events, natural disease and terror attacks. Some contend that the refugee crisis in the Middle East is somewhat brought about by climate problems.

8.4.5. Agricultural Monoculture

Monoculture is the evacuation of most native, wild, or biodiverse landscapes to farm a single species of livestock or plant a single crop.

Most farmland is this way even though, in the modern era, steps are being taken to guarantee that a landscape isn't excessively restricted. In regions where there is a long history of farming, fields are divided with hedgerows and dotted with trees to guarantee as varied a landscape as could be expected under the circumstances.

Progressively, significant cash crops in the emerging world are diversifying, for example, growing coffee underneath tree canopies. This is designated as 'intercropping.' Few proofs recommend that coffee plants are more gainful for this, however, it is additionally useful for biodiversity.

The current large issue with agrarian monoculture is palm oil, regularly added to our toiletries and even food. Foliage is cut down to clear a path for this cash crop, decimating landscapes and creating specific problems for orangutans which are currently threatened species.

8.4.6. Illegal Hunting

In spite of stringent international laws on hunting major games and protecting species traditionally hunted, for example, rhino, and elephant, the illicit pet trade and ivory trades keep on being problematic for international law implementation whose job it is to secure threatened and endangered species.

We have already observed the virtual extinction of the Northern White Rhino because of historic overhunting and an absence of cooperation and coherence from the bodies that could have made the endurance of the Northern White Rhino as successful as its southern cousin.

The black-market trade is additionally an issue. Few Traditional Chinese Medicines utilize rhino horn for relief of rheumatism and apparent painkilling properties. Ivory is likewise on sale in numerous Chinese markets and some in the developed world where checking is slack.

Unlawful hunting that prompts extinction of one species isn't restricted to that species yet has a thump on impacts for others and at times equally threatened species.

8.5. TECHNOLOGY-BASED WILDLIFE IN SITU CONSERVATION

8.5.1. Bio-Logging and Biotelemetry

Bio-logging and biotelemetry have various types of collecting information, however, both involve the checking of behavioral, environmental or physiological information of organisms that are hard to notice or usually otherwise unattainable.

Bio-logging technology stores and records the data in an animal-borne gadget (archival logger) and once the logger is retrieved, the data is downloaded, not at all like the biotelemetry technology, which sends the data to a receiver that is issued from the gadget carried by the animal.

For example, Block expresses that the integration of environmental information with animal gathered information has been less complex with the utilization of emerging electronic tagging and remote sensing satellites, which give a rapid and more precise environment sampling and higher resolutions of global views.

New methodologies utilizing the two technologies are changing the ability to direct ecosystem scale science and to improve the ability of researchers to investigate unanswered ecological questions.

The technology of Global Positioning System (GPS) permits researchers to get exact patterns of movement of an animal through GPS telemetry where the location of an animal and its distance to survey sites can be evaluated.

Figure 8.5. Conservation drones.

Source: Image by Wikipedia.

Such innovation has assisted with identifying, for instance, the utilization of unpredicted habitats, to investigate the social elements of reintroduced species, and to explore unfamiliar historical attributes of threatened species.

Animal borne technology, alluded to as animal-borne video and environmental data collection systems—AVEDs, collects high-resolution datasets that can quantify the animals' physiology, conduct, community interactions, demographics, cooperation, and the environment of the animal inhabits.

A huge variety of these sensor types to gather information on wild animals' external and internal states have been bundled into lightweight units.

For instance, in birds, lightweight geolocators or satellite transmitters have permitted the empirical reconstruction of the migratory routes and wintering zones for small and large birds, which can provide opportunities to test predictions about migration strategies.

Animal borne gadgets are likewise convenient for testing hypotheses about drivers of habitat use. For instance, in the Southern Ocean, a study on southern elephant seal (*Mirounga leonine*), which looked at the geographic distribution of core foraging regions and conduct and evaluated the overall nature of the habitats regionally, exhibits clear advantages of utilizing satellite tracking systems and their assistance to understand more about the animal's reaction to diverse environmental conditions and populace viability. This data is indispensable for creating conservation-oriented management actions.

8.5.2. Camera Traps

Camera traps are remote gadgets outfitted with an infrared or motion sensor that automatically records videos or pictures. They have become a significant tool for wildlife research. The diminishing cost gives scientists extra opportunities to reach and monitor a bigger number of wildlife populaces.

Conventional methodologies, for example, capture, visual, and trapping strategies, can be labor intensive and can require hundreds or thousands of person-hours. Though, camera traps can multiply the number of observers and make them more cost proficient.

The utilization of this innovation has expanded to address questions of species' distribution, population densities, activity patterns among other questions.

Camera traps offer an empirical approach to answer numerous queries regarding wildlife other than the estimation or density of animal populaces. Behavioral studies utilizing camera traps, for example, the first-ever done by Gysel and Davis (1956) where they depicted a basic framework to photograph wildlife, assist us in understanding how various species utilize their habitat.

For example, Bauer et al. analyzed the scavenging behavior of puma in California through camera traps and telemetry. While puma is known to be shrewd predators, their outcomes showed that puma is additionally opportunistic scavengers.

A recent study in chimpanzees (*Pan troglodytes*) inspected community demographic changes (deaths, births, immigration,) and community composition (age and sex structure).

The authors found that camera traps permitted for a rationally accurate approximation of demographic composition and variation among and within social groups. They additionally feature that such innovation may give more precise and accurate measures of fine-scale group abundance.

8.5.3. Additional Technologies

The emerging technology of synthetic biology is expanding swiftly and presently applied to conservation. This field is equipped for altering natural genomes in an extremely precise manner, through deleting a target gene and/or inserting a synthetic one (CRISPR/Cas9 technology), which can carry the viability of genetic modification to another level. A few instances of the principle of conservation issues with potential solutions through the use of synthetic biology are as per the following:

1. Habitat conversion by modifying or creating microorganisms that devour hydrocarbons so as to tidy up oil spills or by utilizing frameworks to create man-made palm oil thus diminishing tropical forests modification.
2. Overexploitation, overexploitation, where the creation of materials that can substitute deep-sea shark squalene or rhino horn ivory.
3. Invasive species, where the utilization of chromosome modifications and quality drives to stop proliferation in these species. The latter is yet connected with moral, ethical, and esthetic issues in which Piaggio and partners call for a risk-assessment framework and robust decision-making in the application of synthetic biology to conservation concerns.

Track plates offer a further productive technique to detect wildlife, and they have been utilized in a variety of approaches to monitor various animal varieties. Initially, track plates were created to monitor rodents' abundance and were in this manner adapted for use with carnivores.

Thinking back to the 1980s, such devices generally contained an aluminum plate in a plywood box, and typically, a bait was put close to the back of the container. The negative track impressions were made after the underlying plate surface was uncovered, when the animal's foot removed soot.

Different tracks of mammals were and are made by utilizing, for example, carbon black or contact paper, ink-covered tiles, smoked kymograph paper sand, mineral oil blend and dispersed printer toner. Track plates are viewed as reliable and economical gadgets that can give robust measurements of animals' abundance.

For instance, Connors, and coworkers, utilized track plates to gauge abundance and local predation risk made by white-footed mouse

(*Peromyscus leucopus*) scrounging activity, and the reason that such gadgets were a dependable method for measuring the local risk of attack by terrestrial mammals without considerably changing the spatial distribution of risk.

A latest study by Smith and coworkers affirms that a well-designed trap to enclose the track plate can be genuinely economical, nonintrusive, and a simple monitoring device. They explicitly looked at whether reproducing the phenology of a generalist predator was related to human reactions to environmental change.

For this, they evaluated the seasonal abundance of small mammals utilizing presence or abundance information gathered from track plates, along with motion-activated trail cameras to acquire visual authentication of the character of small mammals visiting traps.

Environmental conditions can be different because of new extreme temperature and precipitation patterns and novel accumulations and interaction species because of the human-assisted spread of exotic species. Considering this, Wood et al. express that such environmental changes call for conservation to turn out to be progressively prescient.

The Development of technology which assists with anticipating key conservation results including animals' distribution, their demographic and physiological states, and their interactions among individuals and species is direly required.

An investigation on Canada (*Lynx canadensis*) evaluated behavioral differences with environmental conditions by building up a multi-scale prediction model of lynx distribution and found within their outcomes that people, in general, tend to utilize more mature, spruce-fir forests than any other structure stage or species.

8.6. TECHNOLOGY-BASED WILDLIFE EX SITU CONSERVATION

The International Union for the Conservation of Nature expresses that ex-situ assortments incorporate "entire plant or animal collections, botanic gardens and zoological parks, wildlife research facilities, and germplasm assortments of wild and tamed taxa."

The world is confronting a disturbing loss of biodiversity, where inflation of extinction rates is essentially determined by human activities. Aquariums and zoos have taken distinctive conservation activities to relieve dangers to species and their extinction in the wild.

They have added to the actual improvement in IUCN Red List status of species through hostage reproducing and reintroduction conservation measures. Although, their contribution to the conservation of species goes farther than that.

8.6.1. Digital Technology

The utilization of technology in educational settings, for example, handling computer tour guides and guidebooks in tourist destinations or museums, is turning out to be increasingly common these days. Zoological parks and aquaria establishments have since long utilized technology to promote conservation education.

The increase of digital technologies offers the public a more meaningful animal experience while building a higher enthusiasm in conservation campaigns, educational activities and in conservation itself.

At exhibits, interactive computers show information and short films about a specific species, affecting guests' consciousness of conservation issues and conservation-related behavior. A few organizations permit and welcome guests to take immediate conservation action on an issue of their choice by directly contributing money.

Live web cameras managed by zoos show recordings of the animals at the zoo on websites, which are accessible to the general public.

For instance, the Dublin Zoo has live webcams to see live film of the animal from wolves, elephants, penguins, and from the African Savanna region. This innovation unequivocally tries to rouse conservation mindfulness through engaging encounters, which bring people and animals together.

Further uses of innovation in hostage situations, for example, animal conservation and animal behavior, have the aim to increase animal welfare and to aid scientific research in numerous areas.

For example, animal cognition research, which benefits essentially from the utilization of technology, can be a successful method to assess the behavior, mood, and welfare of zoo-housed animals. In the latest study, scientists estimated anxiety responses to unpredictable, repeated, and noisy occasions on simple cognitive tasks in three diverse primate species:

- Chimpanzees (*Pan troglodytes*);
- Japanese macaques*(Macaca fuscata*); and
- Western lowland gorillas (*Gorilla gorilla*).

This sort of examination of the animals' affective or subjective experience is vital to comprehend about the animal's state of wellbeing, which can be essential to their endurance and reproduction.

Another application in welfare research is to give the captive animals with more control over their ecological improvement and surroundings. Control of environmental components, for example, access to open-air areas or to privacy, temperature, sounds, and with non-technological articles can be one of the keys to improving their welfare.

8.6.2. Technology Devices

Over the last 10 years, the utilization of animal attached technology by specialists in zoos has expanded remarkably. Behavioral and welfare researchers can monitor locomotor activity, health issues, movement patterns, postural behaviors, track use of space and a range of behaviors indicative of positive welfare and thus comprehend a bit more the animal's view of its social and physical environment.

Gadgets, for example, GPS are applied in zoos to analyze queries regarding habitat use, patterns of movement and activity levels. This technology has additionally been utilized in surveying relationships among animals.

An investigation on African elephants gathered GPS coordinates to ascertain the average distances between individuals with the aim of determining the social structure of individuals to potentially improve management in determining proper group settings to guarantee the well-being of the individual and group.

Accelerometers are additionally utilized in zoo animals to routinely monitor baseline patterns of conduct and to recognize indications of disease or discomfort.

Information on an animal's conduct through these sorts of gadgets on captive animals can educate and add to the species conservation and management. Specialists outline the benefit of gathering information from captive animals.

For example, technological devices are generally adjusted in calibrated in captive animals before utilizing them in wild counterparts, this includes time-synchronizing behavioral observations with the related gadget readings.

8.7. CONCLUSION

The wildlife conservation practices, if followed properly, can help put a hold on the depletion of the wildlife across the globe. This can in turn help to better the future of the wildlife and the ecology and can also help in saving the forests.

The conservation practices need to be integrated with each other. This can be done by integrating the internal conservatory practices with the external ones. There are some general areas of concern in this regard, which, also deals with the future of wildlife. These are crimes against wildlife, conservation of energy, security of food and water for the animals and the aspect of illegal hunting.

REFERENCES

1. Hundal, S. (2004) *Wildlife Conservation Strategies and Management in India: An Overview*. [ebook] Available at: https://www.arlis.org/docs/vol1/69415913/hundal_edited_final_march_10.pdf (accessed on 21 August 2020).
2. Mason, M. (2018) *Conservation: History and Future of Conservationism | Environmentalscience.Org*. [online] Environmentalscience.org. Available at: https://www.environmentalscience.org/conservation (accessed on 21 August 2020).
3. Nasi, R., Brown, D., Wilkie, D., Bennett, E., Tutin, C., Van, T. G., & Christophersen, T., (2008). *Conservation and use of Wildlife-Based Resources: The Bush Meat Crisis* (p. 50). Secretariat of the Convention on Biological Diversity, Montreal, and Center for International Forestry Research (CIFOR), Bogor. Technical Series no. 33. Available at: http://re.indiaenvironmentportal.org.in/files/Conservation%20and%20use%20of%20wildlife-based%20resources.pdf (accessed on 21 August 2020).
4. Naz, S., & Hussain, M., (2016). Current trends in wildlife conservation: A review. *International Journal of Fauna and Biological Studies*, [online] *3*(5), 44–48. Available at: https://www.researchgate.net/publication/308610297_Current_trends_in_wildlife_conservation_A_review (accessed on 21 August 2020).
5. Olney, P., (2005). *Building a Future for Wildlife: The World Zoo and Aquarium Conservation Strategy*. [ebook] WAZA Executive Office. Available at: https://www.waza.org/wp-content/uploads/2019/03/wzacs-en.pdf (accessed on 21 August 2020).
6. Pacheco, X. P., (2019). How Technology Can Transform Wildlife Conservation. *Green Technologies to Improve the Environment on Earth*, [online] Available at: https://www.intechopen.com/books/green-technologies-to-improve-the-environment-on-earth/how-technology-can-transform-wildlife-conservation (accessed on 21 August 2020).

INDEX

D

E

F

G

H

I

K

L

M

N

O

P

Z